DIFFerent SHADe

Under the Sign of Nature: Explorations in Ecocriticism

DIFFErent SHaDes OF Green

**AFRICAN LITERATURE, ENVIRONMENTAL
JUSTICE, AND POLITICAL ECOLOGY**

Byron Caminero-Santangelo

University of Virginia Press
Charlottesville and London

University of Virginia Press
© 2014 by the Rector and Visitors of the University of Virginia
All rights reserved
Printed in the United States of America on acid-free paper

First published 2014

9 8 7 6 5 4 3 2 1

Library of Congress Cataloging-in-Publication Data
Caminero-Santangelo, Byron, 1961–
    Different shades of green : African literature, environmental justice, and
political ecology / Byron Caminero-Santangelo.
        pages   cm—(Under the Sign of Nature: Explorations in Ecocriticism)
    Includes bibliographical references and index.
    ISBN 978-0-8139-3605-5 (cloth : acid-free paper)—ISBN 978-0-8139-3606-2
(pbk. : acid-free paper)—ISBN 978-0-8139-3607-9 (e-book)
    1. African literature (English)—History and criticism. 2. Ecocriticism—
Africa. 3. Environmentalism in literature. 4. Ecology—Africa. I. Title.
    PR9340.5.C35 2014
    820.9'96—dc23

                                                           2013041283

In memory of my father,
Gennaro Anthony Santangelo
(1929–2009)

contents

Acknowledgments

OVER THE YEARS, THE UNIVERSITY OF KANSAS GAVE ME the time and financial support needed to write this book. During my tenure as a Resident Fellow at the Hall Center for the Humanities in fall 2009 I developed early sections of chapters 1 and 2. I am also grateful to the Hall Center and its director, Victor Bailey, for the Nature and Culture Seminar, where I received feedback on two chapters and was part of numerous productive discussions. Chapter 4 was facilitated by a fall 2011 sabbatical leave from the University of Kansas. In addition, I benefited from a Keeler Intrauniversity Professorship in the spring of 2010; it helped make the project more interdisciplinary through my collaboration with two colleagues, Chris Brown (geography) and Gregory Cushman (history). Thanks to the University of Kansas General Research Fund for summer support in 2006 and 2008 and to the Kansas African Studies Center (KASC) and the English Department for funding trips to many conferences where I presented various parts of chapters. KASC also funded a colloquium on environmental studies and literary studies in Africa, which generated rich discussion.

I am grateful to the Rachel Carson Centre in Munich for a workshop on nature and environment and, especially, to Anthony Carrigan, Elizabeth DeLoughrey, and Jill Didur for organizing it; feedback from and engagement with an amazing group of scholars helped me shape my ideas on the uses of political ecology. Thanks also to Rhodes University, Trinity College, Garth Myers, and Dan Wylie for opportunities to present overviews of the project as it was nearing completion.

I appreciate the permissions granted to reprint revised versions or excerpted portions of the following essays: "Shifting the Center: A Tradition of Environmental Literary Discourse from Africa," in *Environmental Criticism for the Twenty-First Century*, edited by Stephanie LeMenager, Teresa Shewry, and Kenneth Hiltner (Routledge, 2011); "In Place: Tourism, Cosmopolitan Bioregionalism, and Zakes Mda's *Heart of Redness*," in *Postcolonial Ecologies: Literatures of the Environment*, edited by Elizabeth DeLoughrey and George Handley (Oxford University Press, 2011); introduction and "Never a Final Solution: Nadine

Gordimer and the Environmental Unconscious," in *Environment at the Margins: Literary and Environmental Studies in Africa,* edited by Byron Caminero-Santangelo and Garth Myers (Ohio University Press, 2011).

Many thanks to the friends, colleagues, and editors whose feedback, suggestions, and conversation were crucial in the development of this book: Glenn Adams, Folabo Ajayi-Soyinka, Victor Bailey, Chris Brown, Ali Brox, Lawrence Buell, Erin Conley, Dustin Crowley, Gregory Cushman, Brian Daldorph, Elizabeth DeLoughrey, Dorice Elliott, Stephanie Fitzgerald, Peter Grund, George Handley, Jonathan Highfield, David Hoegberg, Graham Huggan, Stephanie LeMenager, Simon Lewis, Julia Martin, Liz MacGonagle, Antonio Melchor, Jeremy Miller, Brian Mulhern, Peter Ojiambo, Anna Neill, Susie O'Brien, Tejumola Olaniyan, Ellen Satrom, Teresa Shewry, Carol Sickman-Garner, John Tallmadge, Brook Thomas, Don Worster, Laura Wright, Dan Wylie, and Boyd Zenner. I also benefited from lively discussions with wonderful graduate students (too numerous to name here) in and outside the classroom.

There are a few people to whom I am especially indebted. I sincerely thank Clare Echterling for her tireless work, keen eye, and superb organizational skills as we readied the final version of the manuscript. The friendship, feedback, and scholarship of Anthony Vital have been an inspiration throughout the years. Paul Outka enabled me to sharpen my arguments through feedback and exciting debate, and his warmth, humor, and cycling mania helped keep me sane during the final writing stages. I am forever indebted to Rob Nixon for his quick generosity, his invaluable advice, and his extraordinary scholarship.

Garth Myers has been instrumental in the development of the project. Years ago he included me in KASC's US Department of State affiliation grant with the University of Zambia, focused on environmental studies; the work I presented in Lusaka became the kernel for the book. Since then, he has helped me develop into a more interdisciplinary scholar through innumerable conversations and suggestions, feedback on many of the chapters, the cohosting of a colloquium, and the coediting of an edited volume. No less important, Garth's good humor and our long friendship have often given me crucial moments of respite when I most needed them.

As always, very special thanks to my family members for their loving support and encouragement: Beth, Jerry, Whitney, Nicola, and Gabriel Santangelo. Finally, my deep appreciation goes to Marta for all the many ways she has helped me through our years together.

Introduction

IN 1991, LARRY SUMMERS PRODUCED A NOW INFAMOUS memo urging the World Bank to encourage "more migration of the dirty industries to the LDCs" (less-developed countries). Part of his "economic logic" included an assertion that "countries in Africa are vastly *under*-polluted" (qtd. in Harvey, *Justice* 366–67). The continent's positioning in this memo is not surprising. Most obviously, by the standards of neoclassical economics, Africa includes thirty-nine of the fifty least developed nations in the world. According to a certain "logic" (which both Summers and the *Economist* deemed "impeccable"), these countries are the most likely to accept pollution in return for economic growth (Harvey, *Justice* 367). At the same time, although there is a long history of environmental degradation in Africa by imperial capital operating with impunity, this degradation has mostly been rendered invisible to the rest of the world as a result of the continent's extreme marginality both in imperial representation and in the world economic system. In fact, its marginalization makes it a great place to do business; minimal media exposure, images of irredeemable chaos and violence, and national governments made weak by globalization often result in the maximizing of externalized environmental costs and the positioning of Africa as a perfect destination for waste.

From an environmentalist perspective, Robert Kaplan seemed to offer a more enlightened perspective than Summers in his *Atlantic Monthly* article "The Coming Anarchy" (1994); yet his representation too has been subjected to withering analysis. Kaplan claimed that resource scarcity caused by anthropogenic environmental degradation and overpopulation is leading to a dystopian future for the developing world and that this future is already with us in Africa, where we find "the political earth the way it will be a few decades hence" (46). In one sense, Kaplan offered a picture of the continent diametrically opposed to the one assumed by Summers; rather than representing Africa as "*under*-polluted," he claimed that "desertification and deforestation" (tied "to overpopulation") were driving more and more people into cities that were already under intense demographic and social stress (46). These conditions were leading to the rapid proliferation of disease, crime, anarchy, and barbaric violence. Kaplan's summation of his thesis was often cited by policy experts in the Clinton White House and in the Pentagon and seemed to signal a new, more environmentally aware political climate: "It is time to understand 'the environment' for what it is: *the* national security issue of the early twenty-first century" (58). However, in at least one very important way, his representation is similar to Summers's: they both suppress the ways that African environments have been and (especially) are being shaped by global political and economic forces and by the long shadow of colonial development. In Kaplan's article, environmental degradation in Africa results from demographics and lack of proper environmental stewardship: "in Africa and the Third World, man is challenging nature far beyond its limits, and nature is now beginning to take its revenge" (54). Ultimately, the article offers us a form of geographic determinism. In a contemporary rendition of the heart of darkness trope, the horrors Kaplan describes are driven by a lack of adequate cultural and social constraints, Western education, and ingenuity. The West looks on, like Marlow, "cut off from the comprehension of our surroundings . . . as sane men would be before an enthusiastic outbreak in a madhouse" (Conrad 37): "Part of the globe is inhabited by Hegel's and Fukuyama's Last Man, healthy, well fed, and pampered by technology. The other, larger, part is inhabited by Hobbes's First Man, condemned to a life that is 'poor, nasty, brutish, and short.' Although both parts will be threatened by environmental stress, the Last Man will be able to master it; the First Man will not" (Kaplan 60). This naturalizing construction has a similar implication for global capital as does Summers's claim that Africa is "vastly *under*-polluted": it reinforces business practices on the continent by suppressing how they have shaped environmental and social crises, and it cuts off consumers from their historical relationship with these crises.

Like Kaplan and Summers, mainstream Western environmentalism has often occluded environmental damage in Africa and/or its complex historical causes. Because of the association of the continent with wilderness replete with exotic biodiversity and charismatic megafauna, parks or potential parks where one finds the real Africa are often highlighted in the Western environmental imagination while the rest of the continent is ignored. This erasure has often caused mainstream conservationists to overlook environmentally destructive extractive industry in Africa (driven by foreign economic interests) and facilitated the creation of conservation enclaves for tourists from which local communities are evicted and excluded.

Such exclusion is tied to a narrative portraying Africans as lacking the proper environmental sensibility and knowledge to take care of precious biodiversity hot spots and, more generally, suggesting that environmentalist efforts in Africa need to be conceived and led by non-Africans. This narrative, reiterated by Kaplan, has an extremely troubling history. Over the last thirty years, many geographers and environmental historians have argued that traditional Western wisdom about environmental change and conservation in Africa has been a form of colonial discourse that works all the more effectively through claims to its scientific validity and/or its apolitical objectivity.[1] Such wisdom celebrates Western environmental knowledge and denigrates indigenous environmental practice, suggesting that Africans do not understand and abuse their environment and that Western experts (or Africans guided by such experts) need to protect it. Explaining why this "received wisdom" has remained so entrenched, Melissa Leach and Robin Mearns claim that it helps promote "external intervention in the control and use of natural resources" and is driven by "the interests of various actors in development," which "are served by the perpetuation of orthodox views, particularly those regarding the destructive role of local inhabitants" (19–20).

Over the past twenty years, the assumptions underpinning these orthodox views have been undermined. For example, in a "path breaking" study, Fairhead and Leach reveal that "outside experts, guided by the wilderness model of nature and negative preconceptions about African land uses, have 'misread' the African landscape" (Neumann, *Making* 57). Such scholarship emphasizes how constructions of an ideal environmental equilibrium in African environments that human impact disturbs suppress the ways "the biodiversity that conservationist biologists identify and covet might very likely be the product of generations of local management" (Neumann, *Making* 152).[2] As a result, the creation of wilderness enclaves through eviction and exclusion based on claims of local environmentally destructive practice have been not only misguided and socially unjust but also ecologically counterproductive. More generally, the colonial discourse of betterment suggesting that African peoples need to be taught proper land use practice has both contributed to and occluded the causes for ecological disasters since the rise of colonial conservation. This discourse denying Africans' environmentalist agency might be suggestively connected not only to Kaplan's arguments but also to Summers's cost-benefit analysis, which implies that there will be no opposition by Africans to exported pollution (Nixon, *Slow* 2).

Since Summers wrote his missive and Kaplan penned his article, events involving African environmental activists have highlighted the limitations of imperialist, marginalizing representations of African environments and environmentalism. In 1995, the writer Ken Saro-Wiwa was executed for his efforts to mobilize the Ogoni people against the destruction of the Niger Delta by the joint forces of the Nigerian state and international oil. Ten years after Saro-Wiwa's execution, the Nobel committee surprised the world by giving the peace prize to Wangari Maathai for her prominent role in the Green Belt Movement, a grassroots women's organization focused on reversing deforestation and preventing the plundering of Kenya's natural environment. As Rob Nixon has brilliantly argued, both Saro-Wiwa and Maathai effectively drew attention to the human costs of environmental "slow violence" in Africa, brought about in the name of development at the expense of impoverished communities, and challenged the association of environmentalism in Africa with fortress-style wildlife conservation driven by the priorities of affluent nations.

Drawing on a critical perspective informed by political ecology and by the theorizing of global environmental justice, this book examines the relationships among African literary writing, anticolonial struggle, social justice, and environmentalism in Africa.[3] In addition to the work of Saro-Wiwa, Maathai, and other explicitly "environmentalist" authors, it takes into account earlier writing that might be characterized as proto-environmentalist. The book has four interrelated goals: first, to bring into question the assumption that Africa has produced little environmental writing; second, to explore how African literature can challenge dominant Western assumptions regarding African environments and environmentalism and how it can offer powerful counternarratives; third, to interrogate widely accepted definitions of environmental writing and the underlying constructions of nature and conservation embedded in them; and fourth, to explore tensions in global environmental justice, political ecology, and African environmentalist writing by putting literary texts in contrapuntal dialogue.

Chapter 1, "The Nature of Africa," links environmentalism in Africa to the particular shaping of the continent as a region by imperialism. This link helps foreground the intersections among African literary studies, postcolonial ecocriticism, and *regional particularism*. It also highlights why *global* environmental justice discourse and political ecology can be a useful means to frame African environmental writing and to explore its significance for conceptualizing resistance to protean forms of imperial development.

The other three chapters all initially highlight a prominent environmental struggle in Africa and focus on a different geographic scale (region, nation, bioregion): the Green Belt Movement in East Africa, the environmental justice movement in South Africa, and the fight against what Michael Watts calls "petro-capitalism" in the Niger Delta ("Violent" 278). Each environmental movement is linked with texts that have received significant attention from ecocritics and/or environmentalists. In turn, these texts are read in relation to earlier anticolonial writing and, more generally, to writing that has been off the ecocritical radar. The readings emphasize how this writing might be aligned with environmental justice and how the seemingly more centered texts need not be read as the origins or endpoints of environmental thought and representation in Africa. Ultimately, creating a dialogue among the texts disrupts a linear or teleological representation of the formation of African environmental thought and writing.

Chapter 2, "The Nature of African Environmentalism," draws on a legacy of environmental writing from East Africa to explore the implications of anticolonial pastoral tropes and antipastoral themes for struggles against environmental injustice. It focuses on the relationships among Maathai's writing, Okot p'Bitek's poems *Song of Lawino* and *Song of Ocol*, Ngũgĩ wa Thiong'o's novel *A Grain of Wheat*, and Nuruddin Farah's novel *Secrets*. Chapter 3, "The Nature of Justice," reads postapartheid novels together with fiction published before 1980 in order to investigate a tradition of South African environmental justice literary writing. The novels include Alan Paton's *Cry, the Beloved Country*, Bessie Head's *When Rain Clouds Gather*, Zakes Mda's *Heart of Redness*, and Nadine Gordimer's *The Conservationist* and *Get a Life*. The final chapter, "The Nature of Violence," brings Saro-Wiwa's narratives of crisis and resistance, *Genocide in Nigeria* and *A Month and a Day*, into conversation with earlier and later writing from the lower Niger: Chinua Achebe's novel *Arrow of God* and poetry by Tanure Ojaide (*Delta Blues and Home Songs* and *Tales of the Harmattan*) and Ogaga Ifowodo (*The Oil Lamp*). Such dialogue, it argues, can help us look in new ways at the literary project of imagining effective struggle for environmental justice in the Delta region, in Nigeria, and in Africa.

1 THE Nature OF Africa

AFRICAN ENVIRONMENTAL WRITING TENDS TO PRIORI-
tize social justice; lived environments; livelihoods; and/or the relationships among environmental practice, representations of nature, power,
and privilege. As a result, it would perhaps be considered inadequately
concerned with "the value of nature in and of itself" (Heise, "Hitchhiker's" 507) and inadequately "ecocentric" (Buell, *Environmental* 21) for
an ecocriticism shaped by mainstream environmental discourse, origi-
nating and centered in the West, which separates nature and its defense
from systemic inequality among humans. Such discourse often implies
that the closer one gets to the truths of ecology and to appreciation and
care for nature, the more one escapes the influence of socioeconomic
interests and becomes a true environmentalist with nature as a constit-
uency. This perspective cannot be separated from notions of objective
representation and forms of desire (for the "freedom of the wild") associated with relatively privileged positions shaped by four hundred years
of European imperialism.

In contrast, viewed from a framework stemming from political ecology and studies of global environmental justice, the notion that African
writing lacking an apparent "ecocentric" focus might be environmental
becomes substantially less outlandish. Such a framework does not posit a
nature that is free (materially *or* conceptually) from mediation by social
struggle, and it undermines stable definitions of environmental threat
and conservation. Concerns with environmental policy are couched in
terms of their connections with economic inequality, social justice, and
political rights and in terms of how they impact the lives—the homes,
livelihoods, and health—of the impoverished and disenfranchised. At
the same time, a critical framework shaped by political ecology and environmental justice remains attuned to what Nancy Peluso and Michael
Watts refer to as "the causal powers inherent in Nature itself" and to the
dangers of rapid, human-induced environmental change for those on
the losing end of development (25). Ultimately, such a framework ties environmental projects with issues of oppression and liberation. Through
its critical lens, texts that do not prioritize the observation of nature or

that only reference environmental change fleetingly or indirectly but that point to the relationship between anticolonial struggle and the fight against environmentally destructive legacies of colonialism can still be considered environmental and can be more important rhetorically in the struggle against ecologically destructive processes than forms of nature and environmentalist writing that suppress histories of empire.

Yet African environmental writing can also be a means to highlight conceptual tensions in global environmental justice theories and political ecology and to consider the significance of different narratives for addressing these tensions. Such critical work draws attention to the potential relevance of literary studies, political ecology, and environmental justice activism for one another. If political ecology offers an extremely useful framework for approaching questions of politics and environment in literary studies, frameworks drawn from the study of literature can offer political ecologists ways of thinking about language, genre, and rhetoric that can enrich and possibly complicate their work. Beyond field-specific academic considerations, foregrounding the intersection of African literature and the struggle for environmental justice can help keep literary studies off the sidelines of important discussions about how to address social conflict, environmental change, and resource extraction in Africa. In this sense, the kind of critical work I pursue takes its cue from ecocriticism understood in its broadest terms: as the exploration of "acts of environmental imagination" in order to contribute to "environmentalist efforts" (Buell, *Writing* 1–2). More specifically, it is closely aligned with postcolonial ecocriticism in its emphasis on how dominant notions of what constitutes environmentalist action, thought, and writing have been shaped by situated knowledge and representation, which are assumed by those who generate them to be objective and universal and which have been crucial components of imperialism (past and present) as it has sought to establish consent.[1]

As part of their focus on decentering ecocriticism, postcolonial ecocritics often emphasize the need for regional specificity. They remain wary of occluding difference in new universalizing forms of discourse, even as they are attuned to the need to think globally in order to address current environmental crises.[2] In this sense, work on individual regions that have received scant attention may be as important as broader, global studies for the process of dialogizing ecocriticism. This argument is especially pertinent for Africa, which has been the focus of fewer ecocritical anthologies or full-length studies than, for example, the Caribbean and which arguably has been more marginalized than other regions of the postcolonial world—especially if one looks beyond South Africa.[3]

A regional focus need not result in a provincializing vision, a narrowing of concerns, or its own suppression of difference at smaller scales. While emphasizing regional alterity that cannot be subsumed by a more universal imperial or postcolonial condition, an approach characterized by what I call *postcolonial regional particularism* still challenges imperial discourse's suppression of global entanglement in the representation of difference. Eschewing a hermetic model of region, such an approach draws attention to the ways Africa has been shaped in particular ways as a result of uneven relationships and processes operating at a global scale. However, it also pays close attention to differences within the continent and resists creating a limiting analytic closure through its regional focus. Concerned with the history of global relationships, with the need to interrogate imperial universalizing discourses, and with local alterity, postcolonial regional particularism potentially contributes to a more decentered, globally attuned (if more fraught) ecocriticism.

Postcolonial Ecocriticism and Africa

Initially, ecocriticism developed as a subfield in Anglo-American literary studies.[4] However, in the past ten years, an increasing number of articles, edited collections, special issues of journals, and monographs have focused on the intersection of ecocriticism with postcolonial cultural studies.[5] Such work has been termed *postcolonial ecocriticism.*[6] It often emphasizes the similarities between the two fields of scholarship, in terms of a sense of political commitment, interdisciplinarity, and the interrogation of capitalist development. It also focuses on the need for postcolonial studies to be more cognizant of "environmental factors" in discussions of place and its significance (DeLoughrey, Gosson, and Handley 5). Even more emphatically, postcolonial ecocritics seek to decenter ecocriticism, both by including more postcolonial texts in ecocriticism and by arguing that postcolonial literature and theory can transform ecocriticism through increased attention to imperial contexts.

Almost all theorists working to develop postcolonial ecocriticism have noted tensions between postcolonialism and what has been called "first-wave ecocriticism" (Buell, *Future* 8) or "American ecocriticism" (DeLoughrey and Handley). Following what is chastised as "the environmentalism of the affluent," first-wave ecocritics favor literary representations that focus on knowing, appreciating, identifying with, and protecting nature in a relatively pure state and/or on "natural" forms of belonging. First-wave ecocriticism has the tendency to erase histories of indigenous peoples, of colonial conquest, and of migrations that disrupted notions of wilderness and rooted dwelling. In his groundbreaking article "Environmentalism and Postcolonialism" Rob Nixon notes the many ways that ecocriticism's "dominant paradigms of wilderness and Jeffersonian agrarianism" all too easily suppress histories of indigenous peoples and the shaping of places by transnational forces (239). In contrast, postcolonial ecocritics attempt to historicize nature (while putting nature back into history) in order to disrupt the naturalization of geographical identities and conditions that have been shaped by imperialism.

Ecocritics and postcolonialists share the goal of decentering the subject; however, for first-wave ecocritics, the subject is decentered in respect to "the nonhuman world" but not in respect to "human others" (Heise, "Hitchhiker's" 507). In moving beyond the human-nature dichotomy, the subject is freed from the differentiation imposed by history and culture into a universal natural condition. The result is a recentered "normative ecological subject" whose ideas and experiences of nature are rendered objective and true, unmediated by language, culture, or social relations (DeLoughrey and Handley 20). Unsurprisingly, first-wave ecocritics often castigate poststructuralism and historical materialism for their skepticism regarding claims to be able to represent nature in ways that escape political positionality. They embrace mimetic approaches to environmental representation, with a focus on the ways literary writing might break through culturally and politically inflected constructions of the environment to achieve a clear, unmediated reflection of the natural world and to give voice to nature. When combined with first-wave ecocritics' valorization of ecology, this position can lead to an uncritical approach to Western science and its claims of scientific objectivity. For the postcolonial critic, a theoretical stance that denies that all modes of knowledge production entail "institutionalized ways of seeing with *histories*" is extremely problematic (O'Brien, "Back" 187). For example, it can unwittingly justify the violence done to indigenous peoples, cultures, forms of knowledge, and places through an imperialism working in the name of objective science.

Efforts to make ecocriticism more responsive to colonial history and to cultural difference have been central to postcolonial ecocriticism. Focused on undermining colonialism's drive for "an unmediated possession of the world," postcolonial theory highlights "the dangers of heeding claims by *any* cultural structures . . . to reflect the world transparently" (O'Brien, "Back" 194). In other words, a *postcolonial* ecocriticism will emphasize the ways that representations of the "material world ('nature')" are situated and "mediated by culture and society" (Vital, "Toward" 90). Both Susie O'Brien and Anthony Vital are especially concerned that ecocritics recognize "the historicity of ecology as modern science," including both its roots in colonial history and its more contemporary universalizing and potentially colonizing impulses (Vital, "Toward" 90). The goal is not to erase ecology's counterhegemonic and even anticolonial potential, but to note how ecology (as discourse) has been rendered ambivalent through its history.

As O'Brien's and Vital's comments suggest, postcolonial ecocritics often explore how "discourses of nature and the environment have been shaped by the history of empire" (DeLoughrey and Handley 10). From the eighteenth century, the idea that nature could be mastered through scientific knowledge had a mutually enabling relationship with the colonial project (Drayton); in particular, Linnaean classification both developed through the collection of unknown species by imperial explorers and encouraged imperial expansion (Pratt; Stepan).[7] Creating "a new planetary consciousness" underpinning "modern Eurocentrism," science instrumentalized the natural world in ways that grounded colonial development (Pratt 15). Thus, for example, it was believed that "agriculture could reclaim wastelands and make barbarous peoples civilized" if it was guided by scientific, "rational" planning (Adams, "Nature" 27). Ecology and conservation also had a close relationship with colonialism (Griffiths; Grove). Richard Grove claims that they were generated in the colonized world (rather than as a response to industrialization in Europe) and that colonial states found conservation advantageous both economically and politically (15). For example, colonial power was enhanced by the notion that, lacking a basis in proper ecological knowledge, "local systems of resource use threatened nature" (Adams, "Nature" 30). Such ideas have by no means been discarded; they often underpin "contemporary thinking on conservation" (Adams, "Nature" 19). For example, efforts to preserve wilderness can still be based on a green imperial romance that historically enabled colonial dispossession through images of pure, untouched natural landscapes in need of protection and, in the process, reinforce new forms of imperialism (Curtin 25; DeLoughrey and Handley 12).

By emphasizing "the historical process of nature's mobility, transplantation, and consumption," postcolonial ecocritics also use environmental history to historicize nature and disrupt discourses of place and belonging that naturalize social relationships (DeLoughrey and Handley 13). They have paid particular attention to Alfred Crosby's concept of "ecological imperialism," in which the exchange of plants, animals, and pathogens during the age of empire sparked massive ecological change in the non-European world and enabled conquest; thus, for example, pathogens "decimated local populations and laid lands open to the blind, cruel but ultimately profitable legal fiction of *terra nullius*" (Adams, "Nature" 20). More generally, of course, ecosystems were transformed rapidly and dramatically as colonists restructured nature and relationships with it for the sake of economic productivity and of their own enjoyment.

In many ways, postcolonial ecocritical work can be linked with what Buell refers to as "second-wave" ecocriticism, which focuses on the positionality of environmental representation and knowledge and which has, as a result, expanded ecocriticism and embraced cross-cultural dialogue (*Future*). However, postcolonial ecocriticism brings a focus both on global imperial contexts and on parts of the world often elided even by second-wave ecocritics, whose expertise remains predominately in American and British literature. Elizabeth DeLoughrey and George Handley reject the very notion of the different waves of ecocriticism, claiming that it "configure[s] postcolonial concerns and methodologies to be secondary developments, a 'second wave' to an unmarked American origin," and marginalizes critiques of predominant American environmentalist assumptions "articulated by indigenous, ecofeminist, ecosocialist, and environmental justice scholars and activists" that often "predate later, more mainstream forms" of ecocriticism (14, 9). In many ways, this study is aligned with their efforts to decenter ecocriticism by reconfiguring "definitions and genealogies"; it explores how African writers and activists have "contributed to an ecological imaginary and discourse of activism and sovereignty" that are not simply derived from American and European environmentalism, and it is especially focused on a tradition of anticolonial writing that will challenge universalizing and "dominant forms of environmental discourse" (DeLoughrey and Handley 8, 14). However, Buell's notion of first- and second-wave ecocriticism remains useful if we think primarily in terms of *literary* studies rather than in terms of "ecological thought" or "environmental discourse" across other disciplines in the humanities and social sciences (which is primarily what DeLoughrey and Handley seem to have in mind); in particular, before 2000 there was little literary criticism that developed out of an environmental justice framework and almost none produced by postcolonial scholars.

In one of the first published discussions of African literature and ecocriticism, William Slaymaker argued "that global ecocritical responses to what is happening to the earth have had an almost imperceptible African echo" and called for both African writers and critics to embrace what he saw as a global ecocritical movement (138). If one only applies first-wave criteria, there has certainly been little ecocritical literary writing from Africa. African writers have primarily addressed pressing political and social issues in colonial and postcolonial Africa. Concomitantly, in terms of environmental representation, these writers are concerned with lived environments, the social implications of environmental change, and the relationships between representations of nature and power. Certainly this is evident in even a cursory reflection on, say, the way the environment figures into the works of Chinua Achebe, Ngũgĩ wa Thiong'o, Nuruddin Farah, Bessie Head, Nadine Gordimer, and even Ken Saro-Wiwa and Wangari Maathai. These writers do not focus on nature in a supposedly pure state and its preservation.

In contrast with such writing, the practice of conservation in Africa has often been underpinned by ideas about a pristine nature that is threatened by *indigenous* environmental practice and in need of protection by those from the West with proper environmental sensibility. Erased by such a narrative are the extensive, intertwined history of nature and culture in Africa and the *creation* of spaces of wilderness through the forced removal of those with long histories of inhabitation. Further, the focus on nature and its preservation in the context of Africa can shift attention from social problems or make such problems secondary to conservation (especially of fauna). In this context, an ecocriticism based on principles from the environmentalism of the affluent will not find much traction in African literary studies.[8]

These arguments are similar to those offered by scholars theorizing postcolonial ecocriticism. However, they also point to the need to take into account the specificity of cultural, historical, and material contexts in Africa; the ways that modernity has shaped Africa; and the kinds of local responses that have been engendered. Discussion of such contexts gestures toward possible ways that Africa might be thought of differently in terms of environment. In the Western imagination, Africa has been and still is framed as a singularity constituted by absence—of time, civilization, or humanity—and this image has served to legitimate the exploitation of places and peoples in Africa. Given the history of this representation, as well as the continent's heterogeneity, it is tempting to dismiss any representation of "Africa" as a place as a fantasy—and a dangerous one at that. As Achille Mbembe proclaims, "There is no description of Africa that does not involve destructive and mendacious functions" (242).

However, this constructed geographical category of "Africa" has also taken on its own reality as a result of history. We cannot ignore, James Ferguson claims, the ways that the imaginary category has been accepted as real and Africa has become what he refers to as a "place-in-the-world," with the term *place* referring not only to "a location in space" but also to a "socially meaningful, only too real, and forcefully imposed" position in "a categorical system" (the "world") of "countries and geographical regions." That Africa is such a "place" is made "easily visible if we notice how fantasies of a categorical 'Africa' . . . and 'real' political-economic processes on the continent are interrelated" (*Global* 6–7). "Africa" as a category may be a phantasm of colonial discourse, but imperialism past and present has also brought this phantasm to life. Africa has become different from the rest of the globe, but this difference can only be understood properly in terms of a history of unequal global economic and political connections feeding off of and giving reality to an assigned geographical position.

Attuned to the significance of Africa's place in the world for environmental change and governance on the continent, this project reflects on what connections, differences, issues, challenges, and opportunities for action are highlighted by focusing on the continent as a particular (and a particularly marginalized) region. However, it also remains concerned with differences within Africa. In this sense, what I term *post-colonial regional particularism* is provisional, always recognizing the need to think about connection and difference across and within scales.

Such an approach is not exactly aligned with the notion of regional ecocriticisms (e.g., an African ecocriticism). The use of particular geographical scales to develop theoretical frameworks can too easily contribute to the hardening of boundaries and the deemphasizing of other scales. (The possibility of nationally defined ecocriticisms brings this fear even more sharply into focus.) Rather, postcolonial particularism is related to a postcolonial ecocriticism concerned not only with bringing the local and global together but also with emphasizing how "each interrupts and distorts the other" (O'Brien, "Articulating" 143). Postcolonial ecocriticism—like ecocriticism and postcolonialism more generally—needs to make connections across cultural and historical difference, but it should also resist suppressing difference in the pursuit of unity. In this sense, the best kind of postcolonial ecocriticism will avoid becoming associated with too narrow a perspective determined by a single theoretical or geographical position. Lawrence Buell has claimed that "ecocriticism gathers itself around a commitment to environmentality from whatever critical vantage point" (*Future* 11). Something similar might be said of postcolonialism, using the term *antiimperialism* instead of *environmentality*. In fact, ecocriticism and postcolonialism run the risk of becoming imperial forms of discourse precisely when they assume the universality of frameworks that are, in actuality, the product of particular geographical and/or social positions—for example, environmentalism of the affluent (in the case of ecocriticism) or poststructuralism (in the case of postcolonialism). If postcolonial ecocriticism allows itself to become tied to overly specific theoretical positions and/or conclusions flawed by the repression of geographical and social difference, it will risk betraying its resistance to colonizing forms of representation and its commitment to true dialogue among different narratives of nature and culture.

Nature, Conservation, and Africa's Place in the World

In discussing Africa as "a place-in-the-world," James Ferguson argues that a key component in the continent's material categorical construction has been "highly selective and spatially encapsulated forms of global connection combined with widespread disconnection and exclusion" (*Global* 14). In other words, Africa's "place" has been constituted by a particular relationality with the rest of the world: that is, specific forms of connection and/or disconnection.

This relationality includes the particular position assigned to Africa in terms of nature. In colonialism's mapping of geographical and temporal difference, Africa has been defined by its embodiment of timeless and dangerous, if awe-inspiring, wilderness. Unlike the Americas, its essence has not been associated with "the garden," defined by accommodation between humanity and nature without predation, but with the essence of monstrous or exotic natural threat. Africa was central in "the darkening of the tropics" that developed from the middle of the nineteenth century, when "the general delight in tropical abundance was giving way to dismay at tropical excess" (Stepan 48). The toll that disease took on Europeans venturing into the African interior was especially important in terms of shaping the "genealogy of the myth of the 'Dark Continent'" (Brantlinger 173–98). Africans fit into this picture either as human manifestations of Africa's wild, dangerous essence or by being erased from the picture altogether.

Often, the African wilderness is given no intrinsic value; it only has positive meaning when its raw material is transformed and made useful for the sake of development. However, in the colonial imagination, wild Africa can also be magnificent, if still deadly: a place to be appreciated but also feared and respected. If this safari Africa in many ways contradicts the image of Africa as horrible, wild chaos, it still erases all precolonial human history and environmental agency. And it still entails the notion of a stewardship of nature in the name of development that centers the European or Europeanized subject and his interests (Brockington 4).

The image of Africa as wilderness has been central for *how* it has been connected to the rest of the world. This image has, in particular, contributed to the connection by capital to small parts of Africa that can be made useful and/or ordered and to the bypassing of the majority of the continent. Drawing on the distinction, from French colonialism, between "usable/useful Africa and unusable/useless Africa," Ferguson notes that in the present "usable Africa gets secure enclaves—noncontiguous 'useful' bits that are secured, policed, and, in a minimal sense, governed through private and semiprivate means," while the rest ("unusable Africa") is "marginalized from the global economy" (*Global* 39–41). These enclaves enable the bypassing of labor and environmental regulations, even as they protect operations and secure the safety of foreign experts, workers, and tourists. Local communities get little or no benefit from these enclaves, although they are the ones whose resources are being taken or rendered inaccessible and who will have to cope with what Rob Nixon calls the "slow violence" of environmental degradation (*Slow*).

Just as was true with the colonial plundering of Africa, the process described by Ferguson is enabled by the perception of the continent as an irredeemable chaos of disease, violence, and poverty shaped not by history but by a savage spirit that makes any kind of progress impossible. Those spots that have desirable natural resources or beauty are secured, while the rest of the continent is left as part of a supposedly undifferentiated violent, degraded wildness. At the same time, the negative impact of global capital is suppressed by the notion that Africa's problems result from its inherent, degraded character. Political crisis in Africa was accelerated by structural adjustment that resulted in the outsourcing of "more and more of the functions of the state," in the deterioration of "state capacity," and, ultimately, in "less order, less peace, and less security" (Ferguson, *Global* 38–39). In turn, extractive industries are not hampered but actually aided by this breakdown precisely because there are fewer checks on their operation and fewer expenses. In a final twist, extractive processes often have socially and environmentally disastrous consequences, but foreign capital's responsibility gets hidden from view precisely by the notion that Africa is too wild and chaotic to understand or develop successfully. Thus, what Achebe called Africa's "image" is not only reinforced by mounting crises but also plays a central role in those crises as it naturalizes the continent's place in the world and obscures the role that imperial exploitation and international entanglements have played in shaping that place.

When Achebe invoked Africa's "image," he did so in the context of his well-known indictment of Conrad's *Heart of Darkness*. Marlow's particular depiction of the continent as irredeemable wilderness has, of course, become a kind of ur-text for which the very title of the novella has become a shorthand. He does condemn the effects of the colonial plundering of Africa's natural resources, including, possibly, the environmental degradation it caused. For example, he indicts the "sordid buccaneers" of the "Eldorado expedition" who "tear treasure out of the bowels of the land . . . with no more moral purpose at the back of it than there is in burglars breaking into a safe" (Conrad 32–33). Yet he still repeats the script of Africa as the epitome of savage, hostile (if horribly awe-inspiring) nature that can only be given positive meaning if it can be subdued, transformed, and made useful by European institutions and that, ultimately, will defeat (in fact, mock) such efforts. The farther he goes into the continent, the more he sees a primordial forest where nature takes on its most unrestrained and dangerous form: "We are accustomed to look upon the shackled form of a conquered monster, but there—there you could look at a thing monstrous and free" (37). Africa's uniqueness, its special place in the world, results from its being the most pure manifestation of a wild spirit challenging the expectations and proportions formed by civilization.

The Africans, when in their proper state, are the human embodiment of this spirit. In describing Kurtz's African mistress, Marlow proclaims, "the immense wilderness, the colossal body of the fecund and mysterious life seemed to look at her, pensive, as though it had been looking at the image of its own tenebrous and passionate soul" (60). In the face of this "fecund," feverish spirit, both alluring and deadly, the right kind of men must try to do the work of civilization; they must, Marlow proclaims, subdue and transform nature. Such work, which requires resisting the wild's allure, is embodied in the training of Marlow's crew. It does not make them fully human, but it does give them some individuation, some meaning, apart from the homogeneity of that "tenebrous and passionate soul." For example, Marlow's native helmsman "was no more account than a grain of sand in a black Sahara," except that he had been made useful, was "an instrument" (51). For Marlow, Africa and the colonial project can be redeemed through such work. However, he finds that the darkness cannot be subdued; it will corrupt and make a mockery of all efforts to transform Africa and make Africans civilized.

Contemporary representations by travel writers and foreign correspondents persistently echo *Heart of Darkness*. For example, in a widely read and touted account of over forty years of reporting from Africa, *The Shadow of the Sun,* the Polish journalist Ryszard Kapuscinski obsessively personifies African nature as an inconceivably powerful, predatory spirit: "Nature on this continent strikes such monstrous and aggressive poses, dons such vengeful and fearsome masks, sets such traps and ambushes, that man lives with a constant sense of anxiety about tomorrow, in unabating uncertainty and dread. Everything here appears in an inflated, unbridled, hysterically exaggerated form" (317). Among the many "poses" and "masks" of this (singular) African nature are a "frenzy" of flora; the "terrifying, monstrous, cosmic power" of its fauna; its "deadly" disease; and its indescribable heat (20, 48, 53). Such description, combined with an emphasis on the "discipline and order" of "biol-

ogy in the temperate zones," underpins Kapuscinski's geographic determinism, in which "each race is grounded in the terrain in which it lives,
in its climate" (20, 5). Since "there is nothing here to temper the relations
between man and nature," since Africans are forever "sparring with
[their] continent's exceptionally hostile nature," they are "culturally,
permanently, structurally incapable of progress, incapable of engender-
ing within themselves the will to transform and evolve" (317, 228). He
may momentarily acknowledge the "endless variety" of Africa, but he
persistently displays "a weakness for the dramatic—and not necessarily original or rigorous—generalization" (Finnegan).[9] On the whole, his
representation of African nature suppresses the shaping of conditions
by contemporary transnational political and economic processes and
naturalizes the often horrific conditions he describes.

Of course, the image of Africa as wilderness does not always foreground the horrific. It has also included the idea of African nature as
magnificently exotic, if still excitingly dangerous. This Africa includes
safaris into atavistic natural time and the creation of pastoral places
where settlers find belonging as stewards of the wild and Africans are
absent, servants, or part of the exotic spectacle. *Out of Africa,* that other
colonial ur-text, peddles this version of the continent. In some ways,
Blixen contradicts Marlow's image of African nature, but it still functions to reinforce colonial power as Europeans become the natural citizens and aristocracy of the "real" Africa, nature's nation. David Spurr
notes that the inconsistency of nature (as a concept) at different moments
in colonial discourse is directly proportional to its "practical function":
to designate difference and assign "hierarchical or ethical value to the
distinctions that inhere in structures of power" (168–69). Something
similar might be said about the more specific concept of "wilderness,"
especially as it has related to Africa's place in the world.

In *Out of Africa,* Blixen proclaims, "The Natives were Africa in flesh
and blood. The tall extinct volcano of Longonot that rises above the
Rift Valley, the broad Mimosa trees along the rivers, the Elephant and
the Giraffe, were not more truly Africa than the Natives were, small
figures in an immense scenery. All were different expressions of one
idea, variations upon the same theme" (20). However, unlike in *Heart
of Darkness,* the spirit of the wild is something to be savored and en-
joyed, at least for those who have not been so transformed by modernity
that they have lost connection with their natural selves. Blixen's settler
creates "a pristine Africa marked by adventure, freedom, and power"

in which "self-discovery, self-definition, and self-discipline are seen as derived from nature" (Knipp 3). In turn, the construction of a sympathetic identification between a natural aristocracy and the wild enables Blixen to legitimate an anachronistic feudal hierarchy in Africa. The Africans, being close to nature, recognize as natural an order in which they are peasants or vassals and Blixen and her friends, who maintain a nonbourgeois sensibility, are their proper lords. Yet what marks a superior aristocratic (European) identity is not simply identification with nature, but the ability to bring the wild and civilized together harmoniously. Thus, Blixen represents herself as able to appreciate and be in touch with nature in a way the bourgeois are not and, at the same time, as having a distance from wild nature that makes her appreciation for it cultured and self-aware, in a way impossible for the natives, who "have no sense or taste for contrasts, the umbilical cord of Nature has, with them, not been quite cut through" (158). By making Africans part of the wilderness and by suggesting their lack of the refined sensibility necessary for appreciation and protection of nature, Blixen positions herself and her kind as the proper owners of the land, able to ensure its well-being.[10] Thus, *Out of Africa* is a conventional pastoral, which, in naturalizing Blixen and her kind's right to possession and power in Africa using stereotypes of African wilderness, erases the actual history of colonial dispossession and exploitation that enabled her to be in the position of "mistress" of a farm (with cheap "squatter" labor) in the first place.[11]

Out of Africa's sensibility could be described as conservationist and, potentially, even environmentalist. The section "From an Immigrant's Notebook," in particular, frequently focuses on the ways that Blixen's experiences in Africa taught her to value nonhuman nature apart from its utilitarian uses. In the short vignette "The Iguana," she vividly represents the beauty of iguanas as they sun themselves: "When, as you approach, they swish away, there is a flash of azure, green and purple over the stones, the colour seems to be standing behind them in the air, like a comet's luminous tail" (246). She then explains that she once "shot" one in order "to make some pretty things from the skin"; however, when she went to where "he was lying dead upon his stone," she saw that "all colour died out of him as in one long sigh, and by the time that I touched him he was . . . like a lump of concrete. It was the live impetuous blood pulsating within the animal, which had radiated out all that glow and splendor" (247). There is a kind of warning in this episode, which echoes throughout the section, concerning the destruction of nature for human use, especially when not driven by pressing need. Blixen particularly abhors the degradation caused by modernity, suggesting in the process that nature has a subjectivity we must respect. She condemns the biological plunder of Africa, in which huge numbers of animals like flamingoes (274–75) and giraffes (288–89) are transported back to Europe for zoos, circuses, and gardens and are destroyed and degraded in the process: "we shall have to find someone badly transgressing against us, before we can in decency ask the Giraffes to forgive us our transgressions against them" (289). Finally, although Blixen does not condemn the hunting done by her younger self and her friends, she subtly marks her maturation in Africa by her movement away from the enjoyment of killing animals toward visual (touristic) consumption of them, as well as, more generally, toward sympathetic appreciation of nonhuman nature and a repudiation of environmental degradation.

In a sense, the environmentalist elements of *Out of Africa* would seem to bring into question colonial narratives of development; yet, on the whole, they (like Blixen's apparent iconoclasm regarding gender, sexuality, and social convention) are by no means in conflict with her construction of her colonial authority. In fact, they help emphasize that she and those like her, in contrast with both Africans and lower-class Europeans, have the sensibility to be the stewards of wild Africa, who will care for it properly. We are frequently reminded, both explicitly and through example, that Africans "have usually very little feeling for animals," and her own elevated sensibility is directly tied with conservation when she mentions (at least twice) her regret that "the whole Ngong Mountain was not enclosed in the Game Reserve" so that the "Buffalo, the Eland and the Rhino" would not be killed off and driven away by the "young Nairobi shop-people [who] ran out into the hills on Sundays, on their motor-cycles, and shot at anything they saw" (34, 6).

Blixen's literary progeny are legion. Another canonical settler pastoral, Elspeth Huxley's memoir "of an African childhood" *The Flame Trees of Thika*, also naturalizes colonial hierarchies and reinforces stereotypes of Africans through representations of relationships between different human groups and the "wild." Huxley often depicts Africans as part of the wilderness, and this identification is again reflected in their similarities to animals. However, *The Flame Trees of Thika* particularly emphasizes that the most "real" Africa is a place where people have never been. In this place, the megafauna "had their wide plains and secret reeded water-holes all to themselves"; there "no man, white or black, had ever set his foot before," and "nature keeps her pure and intricate balance free from the crass destructiveness of man" (182). Africans are written out of this Africa altogether, making it easier to set up the settlers as its proper stewards. In addition, settlers are represented as more admirable in their treatment of animals than the Africans. In the economy of the text, scenes of animal cruelty are among the most powerful examples of "the heart of darkness." However, in sharp contrast with the antipastoral sentiment of Conrad, Huxley evokes the possibility of a pastoral salvation. In her Africa, the darkness is most fully manifested in the Africans themselves rather than in nonhuman nature, which, while savage, is never needlessly so. They can never achieve a moral and spiritual development measured by a profound appreciation of nature and a movement away from the killing and mistreatment of animals. In general, more contemporary manifestations of the safari Africa have only intensified the erasure of Africans from and the sense of their threat to the pastoral refuge (Hughes).

In *Dark Star Safari,* Paul Theroux mocks what he calls "the mythomaniacs of the present day, such as . . . Kuki Gallman," who repackage the clichés peddled by Blixen and Huxley (182). Yet his own conceptualizing of Africa and African nature remains underpinned by related pastoral assumptions. He goes off on his "safari" from Cairo to Cape Town believing he can escape a troublesome modernity "in the heart of the greenest continent": "Travel in the African bush can . . . be a sort of revenge on cellular phones and fax machines, on telephones and the daily paper, on the creepier aspects of globalization" (1, 3). Not surprisingly, he generally finds his pastoral desires thwarted; he has sold himself as the antipastoral travel writer par excellence, following in the footsteps of Conrad rather than Blixen. He offers up what Mary Louise Pratt refers to as the "official metropolitan code of the Third World," associated with decolonization and accelerated modernization, in which deexoticized "places and peoples become . . . repugnant conglomerations of incongruities, asymmetries, perversions, absence, and emptiness" (215). At the same time, during his "safari" he does not exactly give up his construction of a pastoral Africa. The continent can still be "the anti-Europe, the anti-West" he desires: a place where "there was nothing of home," an exoticized "dark star" (117).

In *The Last Train to Zona Verde,* a subsequent travelogue, he describes himself as setting out for Africa with a renewed pastoral hope.

He views his return to "the greenest continent" as "a way of paying respects to the natural world and to the violated Eden of our origins" (11). Again, the opening mostly sets up an antipastoral screed about the horrors of modern Africa. For brief moments of escape and respite, he must turn to the wilderness and the animals. He enjoys his visit to a safari camp in the Okavango Delta, where the super-wealthy ride on the backs of elephants. It departs from the "mass tourism" for "herds of budget-minded tourists," and the owner's ultimate goal of releasing the elephants back into the wild is "worthy": "It was a transcendent experience and an unexpected thrill" (178). However, still troubled by the taming of the elephants, which sacrifices Africa's supposed wildness, he finds his purist moment of escape while observing wildlife in Etosha National Park. The "marvel" of watching "forty elephants" at a wallow is "his reward for visiting Etosha" (190). As he states later, *"Zone Verde,"* which is "a euphemism for the bush," sums "up the Africa that I loved" (228).

Ultimately, Theroux's pastoral fantasies, his nostalgia, and his fears are not as far off from Blixen's as he would like to imagine, and they have similar implications. The threat is a generalized modernity, and the only positive direction for Africa is an impossible return to a timeless wildness that is quickly disappearing. Positioning the continent as the "anti-West" suppresses the ways that it continues to be shaped by contemporary economic processes linked to the West and to his own privilege and position (which, of course, also generate his pastoral desires). The assumptions underpinning his apocalyptic and pastoral imaginary also repress the possibility of effective agency among Africans, including environmental agency. Thus, for example, in *Dark Star Safari* he mentions the process of deforestation in Kenya, epitomized by a deal "made by some Kenyan politicians to sell off hundreds of square miles of protected land in the ancient forest on the mountainside [of Mount Kenya] to loggers and developers," but he completely ignores the decades-long efforts by Wangari Maathai and the Green Belt Movement to fight this process (174). However, he does claim that the abundance "of animals in Kenya's game parks" is due "to the earlier policies of the ubiquitous Richard Leakey" (181).[12] White Africans can apparently effectively protect African nature, but black Africans cannot. The latter leave the Rift Valley, once "a vast, green, empty, curved expanse" with "forests of thorn trees" and abundant game, "overgrazed and deforested and filled with mobs of idle people and masses of ugly huts" (188).

In this sense, Theroux's environmental imaginary, like Blixen's and Huxley's, ties into a colonial conservation discourse about Africa that has driven the creation and protection of wildlife enclaves. This discourse has been constituted by representations of African nature's unique immensity and exoticism; of the "real" Africa as a natural wilderness untouched by culture; of African cultural practices as environmentally destructive; and of effective Western conservation practices based on politically neutral knowledge.[13] Under colonialism, the creation of these parks had significant symbolic resonance, as colonists and white settlers were rhetorically situated as the proper stewards and thus owners of the "real" Africa, while colonial dispossession was justified by the supposed lack of nature appreciation and of proper feeling for animals among Africans. These parks were based on the principles of "fortress conservation," which include "the creation of protected areas, the exclusion of people as residents, the prevention of consumptive use and minimization of other forms of human impact" (Adams and Hulme 10). Enabled by (and reinforcing) tourists' and Western environmentalists' notions of African nature and local environmental practices, colonial-style fortress conservation remains prominent through the operation of wildlife nongovernmental organizations (NGOs) and the tourist industry, working with African governments.[14] Well after the age of national independence, the construction of "advanced" environmental sentiment has perpetuated the dispossession of local peoples by pushing them off their land and refusing them access to natural resources.[15] At the same time, depoliticizing scientific rhetoric has helped to evade the "ethical and political considerations that lie at the heart of policies that ultimately result in land and other resources being targeted for wildlife conservation" (Duffy 2). The outcome has often been ineffective preservation efforts. Violation of conservation laws and regulations can be seen by those living around parks as political acts of resistance against earlier criminal acts of dispossession and the efforts to legitimate that dispossession. Thus, fortress conservation actually endangers biodiversity, since without the cooperation of local populations conservation programs are doomed to failure (Adams and Hulme 13; Adams and McShane xix).

A recognition of this dynamic in the past twenty-five years among not only academics but also officials, policy makers, and conservation groups has initiated "a counter-narrative" of "community conservation" (Adams and Hulme 10). Community conservation focuses on the need to include, rather than exclude, local people in the management of conservation resources not only by having them participate in decisions but also by taking into consideration local development needs. Yet, all too often, community conservation remains shaped by the ideas and policies of the past, by the priorities of Western wildlife NGOs and the tourism industry, and by Africa's place in the world. As a result, the new model echoes the old one in its focus on protected areas, devoid of people, for tourism and the protection of global natural treasures of unique biodiversity. Local people may get some compensation, but "the benefits offered from protected areas rarely meet the losses experienced" (Brockington 3). The inequality in African conservation continues, but "the division now is not race but that conjunction of wealth, class and international influence that determines . . . westerners are most able to consume the benefits of conservation and national elites benefit most from the tourism industry," while rural groups "bear the costs of giving up land and resources" (Brockington 131). Basically, rhetoric on local community empowerment is mismatched with policies that foster elite capture of that empowerment process and instrumentalize "participation" to meet the needs of international donors and Western conservationists. In order to be effective in the long run, community conservation must enhance the livelihoods of the rural poor by giving them "direct access to the resources of protected areas" (Brockington 130). However, such transformation becomes unlikely as long as models of conservation are "grounded more in wealthy westerners' conservation ideals than rural African needs" (Brockington 131).

The creation of wildlife enclaves and the continuation of fortress conservation are also tied to the particular relationship between Africa and globalization described by Ferguson. In that relationship, protected spots in Africa are "'globally' networked," primarily for the benefit of (relatively) wealthy Westerners and a small African elite, while "most of what lies in between" is "simply bypassed or ignored" (Ferguson, *Global* 44). As is true with extraction enclaves, conservation enclaves have been enabled by the policies of structural adjustment. For those shaping these policies, decentralization, democratization, and privatization are typically perceived to go hand in glove, in discursive tactics designed to project an image of local empowerment and enfranchisement; in actual practice, across the continent, disempowerment and disenfranchisement are, ironically, the common result.[16] The ensuing crises in governance spiral together with conflicts over natural resources, while

programs for the decentralized, democratized, privatized, and participatory management of natural resources are often central flashpoints of
governance failures.[17] For the dominant narrative voices on Africa's environmental problems, the crisis points are deemed too grand and important to belong to fictitious nation-states, and so the governance over
them goes to the global scale, with international agencies constructing
notions of "local community participation" that fit their interests (Ferguson, *Global* 42–43).[18] These voices are in tune with an environmentalist discourse that emphasizes the need for "global-level" (as opposed
to national-level) regulation and protection. This discourse is problematic, at least in regard to Africa, precisely because the "global" is not "an
encompassing, overarching spatial *level*" but "a form of point-to-point
connectivity that bypasses and short-circuits all scales based on contin-
guity" (Ferguson, *Global* 42). All that is bypassed, "the generalized des-
titution, the undermining of state authority, and the spread of civil war
on the continent[,] pose[s] fundamental threats to ecosystems that no
system of protected enclaves can mitigate for long" (44). If, on the one
hand, the term *global* enables Western environmental groups to assume
the universality of their perspective and an apolitical ethical stance jus-
tifying their particular interventions, it also contributes to the perpetuation of injustices associated with fortress conservation and, ultimately,
fails to protect wildlife. In this situation, any way forward must include
an awareness of Africa's particular place in the world and the conditions
that have shaped it, including the assumption of an apolitical, universal
language of conservation. The problem is not necessarily in efforts to
protect wildlife in Africa, per se, but in the lack of a will to confront the
structures perpetuating uneven political and economic relationships.

Not surprisingly, for most Africans conservation remains associated with projects driven by white and/or foreign priorities, as well as
by the greed of governing African elites looking to cash in on tourism.
More generally, conditions in Africa make the kind of mainstream environmentalism popular in the United States and the UK unattractive
or, at least, irrelevant to most Africans. Most do not have the resources
that encourage and enable nature appreciation as a leisure pursuit and
that lead to popular movements for protection of "wilderness." Furthermore, the ethics of individual consumption will not be understood by
the overwhelming majority of Africans in the same way as they are by
relatively affluent environmentalists. It may be that global environmental problems—global warming, overfishing of oceans, disposal of toxic

waste—deeply impact many Africans, but most Africans are not primary sources of these problems, nor do many Africans generally benefit from the resource exploitation that engenders them. Finally, if environmentalists in the United States and Britain are often comfortable working within official channels, colonial history and postindependence developments have generally made the hope of introducing change through existing forms of government and corporate institutions problematic in much of Africa.

Where popular environmental movements have developed in Africa, they have been organized around the connection between social injustice and environmental degradation. They have targeted unequal national and global relationships that threaten the health, livelihoods, cultural survival, and general well-being of impoverished majorities or of marginalized minorities. They are often explicitly or implicitly connected to anticolonial struggle, and they are led by figures who are not officially sanctioned experts or who refuse to work within the channels laid down by the government and global capital. In these movements, the approach to what defines pressing environmental problems in Africa, what causes them, and who has the authority to make such determinations can radically depart from models offered by environmentalism among the affluent. This is not to say that these movements have not had substantial, formative relationships with mainstream Western environmental organizations. Ken Saro-Wiwa was encouraged during a visit to Colorado to draw on the environment as a means to organize and mobilize the Movement for the Survival of the Ogoni People (MOSOP), and Wangari Maathai was consistently funded by foreign environmental groups. At the same time, both Saro-Wiwa and Maathai have shaped the direction of organizations such as Greenpeace and the United Nations. However, the movements with which Saro-Wiwa and Maathai are associated had already been launched when they began working with outside groups, and they already had a strong environmental component. In other words, their environmentalism was not derivative, nor was it simply added to a social movement once that movement became influenced by US, European, and supposedly global environmentalism.

African Literature, Global Environmental
(In)justice, and Political Ecology

The authors associated with the struggle for environmental justice in Africa such as Saro-Wiwa and Maathai can be connected with a long tradition of anticolonial literary writing from the continent. In its earlier permutations, such writing often generated counternarratives to the stories of Africa like those offered by Marlow in *Heart of Darkness* and Karen Blixen in *Out of Africa*. It challenged the representation of precolonial Africa as savage wilderness and of the advent of gardenlike order by colonialism. Some of it depicted, or alluded to, beautiful, health-giving, well-managed precolonial indigenous environments and, in the process, projected representations of effective alternative environmental relationships and practices. Much of it focused on the corrupting effects of colonialism, or at least European modernity, including destruction of an effective accommodation between humans and nature. As time went on, such writing increasingly emphasized the perpetuation (if transformation) of colonialism after independence in the consciousness of Africans, in new national institutions, and in the political and economic relationships within the nation and between the nation and imperial powers (old and new). In the face of arrested decolonization, such writing often "wrote back" to earlier nationalist literary discourse, especially in terms of its representations of gender and class.

Read from a perspective associated with global environmental justice and political ecology, much African anticolonial writing, which has previously received little or no ecocritical attention, can be considered environmental or protoenvironmental and interpreted as anticipating or being aligned with texts by activist-writers such as Maathai and Saro-Wiwa. In some of this writing, concerns with environmental change are not central. In all cases, texts foreground the dangers for people of environmental change, particularly in terms of loss of livelihoods, health threats, cultural loss, and decrease in quality of life. Any suggestion of the need for conservation of biodiversity (especially megafauna) is connected with such dangers, and a focus on the value of wilderness preservation is conspicuously absent. Yet if these texts would be considered too anthropocentric in *priority* for a certain kind of Western ecocritical orientation, they often both emphasize the benefits of indigenous epistemologies that decenter human agency and point to destructive effects of relationships with nature based on mastery and instrumentality; such effects are represented as part of a more general degraded condition initiated by colonialism and characterized by increasing inequality, violence, division, and oppression. In addition, the African writers raise the possibility that applying definitions of proper environmental(ist) sensibility based on supposedly ecocentric criteria may very well be another colonizing move working through the guise of apolitical objectivity and universality. Challenging hegemonic narratives (including narratives of ecological identity, environmental change, and conservation), they point to the idea that the projection of a universal subject decentered in relation to nature but not in relation to other humans can all too easily be a means of disavowing the shaping of certain kinds of environmental discourses by positions of privilege and of suppressing the connection between these discourses and the reproduction of imperial relationships.

In this context, the term *environmental justice* has some potential limitations. It still carries some baggage resulting from its association with research in and on the United States and, to a lesser extent, other parts of the West.[19] For example, it suggests a focus on race and the distribution of waste and industrial sites and with who gets what within the cities and regions of the "developed" nation-state rather than with "questions of distribution, disproportionate impact or marginalization extending beyond the borders of the US to encompass people elsewhere" (Walker 34). This kind of environmental justice frame can fail to account for the ways colonial legacies and neocolonialism result in differences between the sociopolitical conditions in postcolonial nations and conditions in the United States. Williams and Mawdsley note how using such a frame can suppress issues regarding the formation of postcolonial states by imperialism, and they reflect on how "the emphases and implicit assumptions of western environmental justice literature . . . may need reconsideration when working in postcolonial contexts" (660). Even for those like Gordon Walker trying to take into account the transformations of environmental justice in different parts of the world, the association between the term and movements in the United States can encourage a narrative of diffusion in which a concern with environmental justice has spread from the center (the United States) to the peripheries.[20]

Yet using *global* to qualify *environmental justice* still potentially marks a departure from some of the nationally bound or centered conceptions of the term, especially if it is viewed as related to the notion of an "environmentalism of the poor." In fact, I do not use the latter term to define my approach for only two reasons. First, its potential emphasis on class might be limiting in the discussion of national and/or transnational inequalities, and second, it might suggest separation rather than connection with environmental justice movements in the United States and Europe (a particular problem in the case of South Africa).

Theories of environmentalism of the poor focus on struggles against environmental injustice within countries of the Global South and entailed by unequal global patterns of distribution. The term includes forms of environmental action that have not necessarily been inspired by or spread from the United States and that may even predate the growth of environmental justice in the United States. The scholars most closely associated with environmentalism of the poor, Martinez-Alier and Ramachandra Guha, use "historical cases of environmental conflict [from the Global South] which were not yet represented in the language of environmentalism" to help "interpret, as environmental conflicts, instances of social conflict today where the actors are still reluctant to call themselves environmentalists" (Martinez-Alier 54). In similar fashion, the reading strategies developed in this project aim to define as *environmental* earlier and more recent African writing that is outside or marginal to dominant notions of environmentalism and, in the process, to challenge hegemonic notions of what that term might mean.

Environmentalism of the poor often leads to a deep suspicion of institutional channels and universalist rhetoric that ignore inequalities within and beyond the nation. This deep suspicion marks its difference from environmentalism of the affluent. In many of its manifestations, the latter works closely with industry and government and ascribes to a "gospel of eco-efficiency" based on the principles of neoliberal development but with an emphasis on sustainability and ecological modernization (Martinez-Alier 5). Even in its more antiestablishment forms, usually as part of "the cult of wilderness," environmentalism of the affluent has often represented a retreat from the political relations among humans and from the inequalities and divisions generated or exacerbated by development (Martinez-Alier 1–2). For example, the deep green movement has all too frequently ignored the need for action outside protection of wilderness and the human suffering resulting from environmental damage (Guha, *Environmentalism* 87). As Rob Nixon trenchantly notes, in assuming "that the United States represented the environmental vanguard and that wilderness preservation—as philosophy and practice—needed to be universalized," wilderness advocates revealed their "purportedly selfless biocentrism" to be "imperially anthropocentric in its America-knows-best development ideology" (*Slow*, 254). With the word *animal* replacing *wilderness,* something similar might be said of all too many animal rights crusaders.

In contrast, in environmentalism of the poor, activists are attuned to the ways that official, universal discourse enables marginalized communities and their places to come under the disciplinary sway of external interests and more privileged groups. They often point to how the association of institutionally sanctioned expert knowledge with the articulation of objective truth all too easily becomes a means of denigrating other forms of knowledge and other voices and, therefore, of silencing dissent, even as the knowledge itself is a vehicle for the perpetuation of social inequalities and injustice. Martinez-Alier emphasizes, in particular, how the instrumental language of the market and of monetary valuation will result in costs visited on the poor and least privileged and benefits going to the wealthy and more privileged (Martinez-Alier 150; Harvey, *Justice* 388). This process entails not just issues of distributive justice (the distribution of environmental goods and bads) or procedural justice (how decisions are made and whose voices are heard) but also what David Schlosberg refers to as justice as recognition. As he argues, the devaluation of some people, their knowledge, and their right to represent themselves is entailed by, indeed often precedes, distributive and procedural injustice.

Of course, activists associated with environmentalism of the poor draw on scientific, legal, and economic language in their rhetoric, but they often give equal or greater priority to other idioms in framing arguments regarding the significance of environmental problems and solutions to them: "the respect for sacredness, the urgency of livelihood, the dignity of human life, the demand for environmental security, the need for food security, the defense of cultural identity, of old languages and of indigenous territorial rights" (Martinez-Alier 150). Such idioms frame conflicts "as clashes of incommensurable value" and, more generally, as contests between "official discourses" and the points of view they suppress or make invisible (Martinez-Alier 150). For example, activists will challenge "the distancing rhetoric of neoliberal 'free market' resource development" and represent in concrete and vivid terms "the human and ecological costs" paid by marginalized communities and peoples in order to maintain the lifestyles of the wealthy and privileged (Nixon, *Slow* 26). In the face of abstracting discourse, these activists must make visible the "convulsive, material effects" of "corporate power" (*Slow* 169). Ultimately, environmentalism of the poor strives to challenge and reimagine development in order to make it more equitable in terms of distribution and more sensitive to the value of human dignity and ecological integrity.

Scholarship on environmentalism of the poor and on environmental justice movements more generally also stresses the challenge of conceptualizing place in its relationship with what is positioned as "outside." These movements are defined precisely by their efforts to give voice to silenced or marginal perspectives as situated in particular communities and to resist universalizing discourse. In this sense, their militant particularisms necessarily work against unified praxis. Yet, if environmental justice movements focus on specific conflicts, inequalities, and ecological change *in place,* they must still consider the need to address processes and systematic structures (national and international political relationships, the movement of global capital, transnational environmental changes) operating at larger, more abstract geographic scales. If they do not, then their goals may very well fail to take into account threats that will make any successes short term or the possibility that such successes will entail a geographical shifting (rather than reduction) of inequity. Geographies of injustice working at different scales must be taken into account if threats to local ecology and lives are to be addressed in a lasting way and/or are not simply to be moved to other places.

Questions of geographic scale point to one of the most important connections between the theorizing of the environmental justice struggle and political ecology. Dissatisfied with apolitical explanations of environmental problems and with the assumption that "better stewardship of the earth" follows improvement in "the ethical register of people," political ecologists seek to address the "complex political, cultural, and social dynamics" of environmental problems (Peet, Robbins, and Watts 24, 10). They have shown a particular concern with what Michael Watts and Richard Peet refer to as "the politics of scale": that is, the political dynamics among different spatial arenas (the body, local community, nation, geopolitical region, space of global capital, etc.) and different temporal registers (4). Political ecologists suggest that analysis focused only or primarily on ecological change as driven by the individual producer/consumer or local cultural and social conditions is inadequate as it fails to take account of unequal political dynamics and of the power of capital working at different but interrelated scales. To this concern with the politics of *geographic* scale Nixon has added a focus on time. Closely examining the politics of temporal scale enables us to interrogate representations of violence as framed by spectacular time in relation to the unspectacular time of slow violence, an attritional "violence of delayed destruction that is dispersed across time and space" and "that is typically not viewed as violence at all" (2).[21]

In terms of ecology, political ecologists insist on the "causal powers" in natural processes (rather than "solely on the overwhelming forces unleashed by capital, state, or technology") and pay attention to actual ecosystems where "trends in the global economy play themselves out in many unexpected ways" (Peluso and Watts 25; Peet, Robbins, and Watts 23). Yet they also often emphasize that "nature and society are internal relations within the dynamics of a larger socio-ecological totality" and question the existence of a "solid, independent 'external nature'" (Harvey, *Cosmopolitanism* 232). As a result, for many political ecologists change cannot be understood in terms of a society's alienation or liberation from "a nature" to which humanity "was once intimately attached" (i.e., away from or toward a more natural society) (Harvey, *Cosmopolitanism* 231). Instead, it must be understood as the movement from one socioecological totality toward a different one (with a different but still thoroughly historicized "nature"). This perspective helps foreground questions of how produced environments are entangled with global capitalism and its inequalities and how socioecological totalities shaped by other economic and social logics might result in what Neil Smith refers to as "the positive production of alternative natures" and the alleviation of injustice (51). In sum, in political ecology (as in postcolonial ecocriticism) the framing of environmental transformation as a problem (or not) is always inflected by social struggle: "Behind every story of environmental crisis . . . is a narrative of political and social control" (Peet, Robbins, and Watts 37).

To varying degrees, the literary texts I discuss also suggest that environmental projects need to be understood as political projects; they emphasize the relationships among such projects in Africa, discourses of nature and conservation, and sociopolitical injustice. In addition, they share political ecology's concern with "the causal powers inherent in nature itself" and with ecological interdependence. They point to the long-term impact of massive socioecological change wrought by imperial modernity and its accompanying lack of humility. For these writers, the decreasing self-determination and threat to survival accompanying such change suggest the need, particularly for those on the losing end of colonial development, for more decentered ways of approaching our relationships with nonhuman nature.

Yet using only the *commonalities* within and between global environmental justice struggles and political ecology as a basis for a discussion of African environmental writing would be clearly limiting and misguided. It would suppress important activist and academic debates and the value of literary texts and analysis as means to address vexing and important issues. It might also suggest an extremely problematic analytic closure. For example, both environmental justice movements and political ecologists struggle with the question of how to think and work across scales without suppressing the particularities of place and evacuating otherness in reductive universal discourse. David Harvey notes that "we have yet to come up with satisfactory or agreed-upon ways" to "'telescope in' the political insights and energies that can be amassed at one scale into political insights and action at another" (*Cosmopolitanism* 229). Literary texts and investigations of relationships among them can serve as important means to confront this conceptual and imaginative challenge. The African literary texts addressed in this book offer varying approaches to the following questions: How is the delimitation of the particular (threatened) place, community, and subject to be conceived? How are their relationships with what is positioned as outside and/or foreign to be represented? How are the dichotomies "indigenous/foreign," "traditional/modern," "local/global," and "inside/outside" to be understood? As a result, putting the texts in dialogue both enriches discussion about them as interventions in activist and academic discourse about praxis and enables critical consideration of different ways of imagining relationships across scale.

Such dialogue also helps position the texts in relation to debates about representations of nature and identity. Political ecologists often challenge the construction of a community as "the natural embodiment of the 'the local'" and "as a unity, as an undifferentiated entity with intrinsic powers, which speaks with a single voice" (Watts and Peet 24). Such work insists that the formulation and reproduction of communities must always be understood in terms of internal dynamics of power and of the interrelationships between local and nonlocal politics. Some studies have even foregrounded "the pitfall of eco-populism in terms of the ways in which environmental movements *create* a sense of community and tradition for political purposes" (Watts and Peet 19). In this sense, political ecology might be read as offering the tools for unearthing the socioecological unconscious of naturalizing, romantic representations of community and resistance in African environmentalist writing. (The term *socioecological unconscious* refers to the complicated relationships entailed by environmental threat or crisis and suppressed in different kinds of representation; political ecology might be defined as the study of these relationships.)

Yet narratives representing supposedly romantic images of indigenous communities as necessarily suspect or useless can also be construed as having a socioecological unconscious. Much activist writing and postcolonial scholarship suggests that in order to understand the significance of such representations with any specificity, we must take account of "the broader historical and political contexts" out of which they emerge and in which they are received (Chrisman 192). In other words, their meaning needs, at least in part, to be understood rhetorically and strategically (Parry; Chrisman). For example, they may be necessary for the mobilizing of collective resistance and in order to overcome the divisive and debilitating effects of (neo)colonialism. To depict such representations as *necessarily* problematic for antiimperial resistance may very well result from potentially universalizing "conceptions of différance" and "the politics of difference" that suppress the value of unity and the possibility "that people may share needs, values, interests that override their differences"; such an approach can all too easily elide the question of when "the promotion of politics of difference [serves] socially emancipatory or disabling ends" (Chrisman 189, 194, 198).

My contrapuntal analysis of African literary texts draws attention to the intersection between these texts and the complex questions naturalizing representations raise for environmental justice struggles and political ecology. More generally, a critical dialogue among various *kinds* of African literary narratives produced at different historical moments and places and engaging with a range of rhetorical situations can help not only explore the varied landscapes of African environmentalist writing and environmentalism in Africa but also resist closure in the imagining of effective ways to move toward a more equitable, sustainable future.

2 THE NATURE OF African Environmentalism

IN HER MEMOIR *UNBOWED*, WANGARI MAATHAI DRAWS on what Lawrence Buell refers to as an "indigene pastoral" in order to give narrative shape to her vision for social and environmental regeneration in Kenya. She begins her story with a childhood memory of her home village of Ihithe in the central highlands, where a beautiful, health-giving, and well-managed natural environment sustained and defined the human community: "I am as much a child of my native soil as I am my father . . . and mother" (4). Culturally, this community was marked by its animism and reverence for nature: "For the Kikuyus, Mount Kenya . . . was a sacred place. Everything good came from it: abundant rains, rivers, streams, clean drinking water. . . . As long as the mountain stood, people believed that God was with them and that they would want for nothing" (5). Maathai erases colonialism's presence from this initial description of her childhood home in the central highlands as a means to emphasize its catastrophic environmental impact resulting from "logging, clear-cutting native forests, establishing plantations of imported trees, hunting wildlife, and undertaking expansive commercial agriculture" and, ultimately, from cultural transformation: "Hallowed landscapes lost their sacredness and were exploited as the local people became insensitive to the destruction, accepting it as a sign of progress" (6). Throughout the rest of her narrative, she repeats this theme, while insisting that redemption can be achieved through the rejuvenation of indigenous cultural values and the struggle against the legacies—especially psychological and ideological legacies—of colonialism.

In using pastoral discourse, Maathai is, of course, following in a long tradition of environmental writing and rhetoric. Despite its many problems, this discourse has been a crucial "ground-condition" for those struggling against the environmental implications of modernity's narrative of development and positing alternatives (Buell, *Environmental* 32). Yet the particular kind of indigene pastoral Maathai uses also connects her narrative with an important tradition in African letters. The story of natural, harmonious precolonial African cultures and of the corrupting impact of colonialism was a prominent aspect of Negritude, which began in the 1930s and remains prominent in African poetry, fiction, and drama. Its purpose is usually explained in terms of counternarrative: the effort to create stories challenging imperial representations in which Africa is defined by negation (the absence of history, development, civilization, etc.) and in which the coming of the heroic European conqueror represents the advent of a proper ordering of (wild) nature. These representations are part of a " 'worlding' of what is today called 'the Third World' " (Spivak, "Rani" 247) in which places and peoples are placed in a hierarchical global order according to their relative progress on a universal temporal timeline of development. In this order, Africa and Africans are always at the bottom, without any history of effective agricultural or environmental practice.

At the same time, the pitfalls of anticolonial pastoral narratives have been imaginatively embodied in antipastoral African literary writing. An antipastoral tradition typically entails "exposing the distance between reality and . . . pastoral convention"; these realities "need not only be economic or social" but can entail "cultural uses of the pastoral that an anti-pastoral text might expose" (Gifford 128). A novel like Nuruddin Farah's *Secrets* mocks naturalizing representations of "traditional" culture and suggests that appeals to authoritative secrets of nature are tied to privilege, the lust for power, and exploitation in the postcolonial family and nation. Part of Farah's attack on the African pastoral literary tradition entails the suggestion that it has contributed to socially devastating ecological degradation by encouraging the instrumentalization and exploitation of biotic communities. However, if *Secrets* highlights possible problems in the political deployment of pastoral narratives, the history of the Green Belt Movement might also suggest drawbacks in antipastoral representation foreclosing such deployment as a liberatory political strategy.

Development and Its Discontents: Okot p'Bitek, Ngũgĩ wa Thiong'o, and Wangari Maathai

The similarities between Maathai's use of anticolonial pastoral tropes and their uses in older, canonical anticolonial literary texts, Okot p'Bitek's *Song of Lawino/Song of Ocol* and Ngũgĩ wa Thiong'o's *A Grain of Wheat,* point to a legacy of environmental writing and rhetoric from East Africa. It may be that in many cases the indigene pastoral has "more to do with reinvention of the non-European world as a mirror-opposite of certain European norms" than with "actual environments" (Buell, *Environmental* 68); however, such narratives, even those from the nationalist period, cannot be lumped together. Ngũgĩ and Okot may not be as focused on environmental degradation or ecological relationships as Maathai, but, like her (and unlike many of the poets of Negritude), they bring attention to actual environmental changes wrought by colonial ideology and policy and to benefits of (relatively) concretely defined indigenous environmental practice and epistemology.

Reading Okot's poetry and Ngũgĩ's novel contrapuntally with *Unbowed* and Maathai's essay collection *The Challenge for Africa* also highlights potential limitations of the use of the anticolonial pastoral. Writing about romantic perspectives on indigenous relationships with nature in African environmental history, William Beinart warns that "arguments rooted in an anti-colonial and sometimes populist or antimodernist discourse can present us with an analytic closure, too neat an inversion" ("African" 284). In his effort to challenge the continued use of homogenizing, dehistoricized binaries (modern/traditional, Western/indigenous, etc.), Beinart points out that precolonial, colonial, and postcolonial African rural communities have been and are in flux, and their environmental knowledge and practices are not outside relationships of power. Citing the well-known example of Great Zimbabwe, along with others, as evidence of unsustainable environmental practices in the precolonial African past, Beinart argues that African practices were by no means always in harmony with local ecology. Among the political implications of idealism in anticolonial pastoral narratives is an oversimplified solution to current environmental problems (a cultural return) and the eliding of structural political and economic relationships at local and global scales. In addition, these idealist narratives are often tied to a nationalist discourse of communal and geographical essences. This discourse can all too easily be enlisted in the cause of national or ethnic chauvinism—the stigmatizing of the "foreign"—and used to legitimate existing inequalities and injustice through appeal to a notion of authenticity. The accusation of cultural inauthenticity against those trying to formulate counterhegemonic representations has been wielded particularly strongly against women in defense of patriarchal power in postcolonial nations.

Many of these same criticisms have been leveled at the trope of the eco-Indigene, which takes its most recognizable form in the "the Ecological Indian: the Native North American as ecologist and conservationist" (Krech 16). This trope is an extension of the "noble savage" stereotype and projects idealized images of indigenous societies harmoniously dwelling with nature in unfallen Edens and embodying an impeccable environmental ethic. The historic veracity of these stereotypes has been questioned by scholars such as Shepard Krech and Andrew Ross, and they have been criticized for circumscribing the agency and humanity of indigenous peoples by making them "mystical (and therefore unrealistic), historically irrelevant, beautifully and tragically vanished," as well as "less culturally developed" than peoples of European ancestry (Sturgeon 76–77). Critics of the trope consider it the product of white liberal guilt and mainstream environmentalism in the West, although it is usually also acknowledged that indigenous activists use it "strategically" (Kolodny; Sturgeon).

Ugandan poet Okot p'Bitek's *Song of Lawino* is among the most influential poems in twentieth-century African literature. First written in Acoli, and then translated by the poet himself, it was revolutionary in its accommodation of the oral tradition. Okot used the language, imagery, forms of address, and rhetoric of evocation of traditional Acoli "songs," and especially songs of abuse, to develop a new kind of long poem (Heron, *Poetry*, introduction; Okpewho 302–5). He also effectively dramatized the tensions between advocates of tradition and of modernity and, as a result, was widely read throughout East Africa. In the "song poem," the eponymous character by turns ridicules, refutes, and beseeches her husband Ocol. The Western-educated Ocol parrots colonial discourse and, in particular, continuously offers up the image of Africa as a savage wilderness with nothing of intrinsic value; he wants to utterly transform the continent by destroying African cultures, environments, and selves and replacing them with their "civilized" European equivalents. He has rejected Lawino, the voice of tradition, as the representative of all he has come to despise. In her song, Lawino strikes back both by emphasizing the value of Acoli culture and local nature and by representing Ocol's attitudes as embodying a form of corruption and illness. In a shorter (later) poem, Ocol responds to Lawino; however, *Song of Ocol* mostly reinforces Lawino's perspective by suggesting that his attitudes can only result in self-hatred and are, in fact, leading Africa toward social decay and environmental destruction. Read together, the two songs offer a pastoral discourse pointing to the artificiality and destruction brought by colonial modernity and calling for the return to traditional ways of life that are closely tied with environmental conditions and that enable self-sufficiency, health, and an authentic existence.

Lawino's primary argument can be summed up by her song's central, repeated proverb: "The pumpkin in the old / homestead / Must not be uprooted" (41). To uproot a pumpkin plant is to destroy the source of a highly valued, nutritious food and to undermine the ability to sustain oneself. Lawino associates the pumpkin with Acoli culture, which she represents as being effective and beautiful precisely because its "roots reach deep / into the soil" (41). She ridicules the idea of replacing Acoli practices with Western ones that are divorced from local conditions and rendered useless, meaningless, ugly, and/or dangerous in Africa. Often, her criticism of Western culture subtly becomes more universal; she not only ridicules its manifestations in Africa but also suggests that it is, in general, less attractive, more abstract, and ultimately less natural—less in touch with the sensual world—than Acoli culture.

However, the "pumpkin" refers not only to an Acoli culture rooted in a local natural environment but also to that environment itself. Accepting the construction of Africa as one vast monstrous wilderness, Ocol conceives of African nature as offering nothing of value, even as he acknowledges no effective accommodation of nature by humans before colonization. As a result, he seeks the utter transformation of African environments, as well as cultures and identities. Lawino refutes Ocol by alluding to a vibrant Acoli agro-ecosystem ("an ecosystem reorganized for agricultural purposes" [Worster 52]), based on indigenous biodiversity, which enables healthy, fulfilling lives. According to Don Worster, subsistence agro-ecosystems developed by traditional small-scale farmers, "despite making major changes in nature, nonetheless preserved much of its diversity and integrity," an achievement that "was a source of social stability, generation following generation" (56).

Lawino references the benefits of such an agro-ecosystem when she discusses Acoli food and food preparation. Ocol claims that "Black people's foods / are primitive" and "dirty" (62). Lawino responds by arguing that knowledge and utilization of local biodiversity embedded in Acoli culture result in effective, even beautiful food preparation and tasty, health-giving food. To demonstrate, she takes Ocol—and the reader—on a kind of culinary tour of her "mother's" house. For example, she discusses "in detail" the different kinds of firewood "stacked right to the roof" (60). She knows "their names / and their leaves / and seeds and barks" and explains the uses to which each one can be put. In this way, she emphasizes the knowledge entailed in Acoli cooking, the control that it allows, and the value of the indigenous flora. She also explains the advantages of serving and storage technologies her family uses, which are based on local environmental products and highlight the freshness and sensual attraction of local foods:

> The white man's plates
> Look beautiful
> But you put millet bread in it
> And cover it up
> For a few minutes—
> The plate is sweating
> And soon the bottom
> Of the bread is wet
> And the whole loaf is cold.
>
> A loaf in a half-gourd
> Returns its heat
> And does not become wet
> In the bottom;
> And the earthen dish
> Keeps the gravy hot
> And the meat steaming. (61)

Through such sharp observations, Lawino builds up an image of the wholesomeness of a way of life rooted in local nature; at the same time, this image points to the value of an ecosystem that underpins that way of life—it too is the pumpkin that should not be uprooted.

The threat of environmental uprooting is most fully represented by Okot in the later poem, *Song of Ocol,* which expands on Lawino's warnings about the dangers of a ruling elite whose outlook and desires have been utterly shaped by colonialism and who want only to take the place of the former colonizers within the new nation. Such a situation results in the desire to destroy anything African and, ultimately, in self-hatred, since the transformation to a fully Europeanized self is impossible: "Mother, mother, / Why, / Why was I born / Black?" (126). An extreme manifestation of Ocol's hysteria, driven by the contradictions of the colonial discourse he parrots (which simultaneously promises and withholds transcendence), is his desire to annihilate the very geography of the continent; he wants to "uproot/Each tree/From the Ituri forest," "blow up/Mount Kilimanjaro," and "divert/The mighty waters/Of the Nile/Into the Indian Ocean" (146).

As a means for Okot to express his vision of the contradictions, self-hatred, and hysteria of a Western-educated ruling class's colonized condition, Ocol's call for the geographic transformation of Africa works well. However, the apocalyptic imagery is both too improbable and too melodramatic to serve as a warning regarding the threat of actual human-induced environmental degradation. More subtle and more convincing in this regard is Ocol's desire to eradicate the existing agro-ecosystem, as well as his recounting of the "progress" he has made toward this goal. Early in his song, Ocol proclaims,

> I see an Old Homestead
> In the valley below
> Huts, granaries . . .
> All in ruins;
>
> . . .
>
> We will plough up
> All the valley,
> Make compost of the Pumpkins
> And the other native vegetables (124)

Later, he makes clear the kind of rural landscape that will replace the one that has been destroyed; the ploughing and composting he references will be part of the establishment of mechanized monocropping on large, privately owned farms. Describing the history of his own property, he notes that when "the tractor first snorted/On these hunting grounds/The natives" were like "squirrels" chased "by the hunter's dog" and proclaims "Africa's wildest bush" to be "now a garden green/With wheat, barley, coffee" (141). Ocol embraces a form of capitalist agriculture that includes the enclosure and privatization of the commons (the "hunting grounds") for the production of cash crops; in the process, the land itself has been turned into a commodity that can be concentrated in the hands of the new political and economic elite running the freshly independent nation: "We have property/And wealth,/We are in power" (Okot p'Bitek 142). The hopes of social justice, of a focus on the well-being of all citizens, are mocked, as the elite continue processes that began under colonialism in Africa. These processes are sanctioned by the language of development, which represents them as turning useless "bush" into a productive "garden."

In her writing, Wangai Maathai has stressed how the kinds of attitudes toward traditional subsistence farming expressed by Ocol, and the concomitant pursuit of cash crop agriculture, have been ecologically and socially disastrous. Lawino's warning not to uproot the pumpkin of traditional agro-ecosystems has gone unheeded, much to the peril of East Africa. As farmers have cleared their land for crops such as tea and coffee, they have contributed significantly to deforestation, leading to soil erosion, loss of fresh water sources, scarcity of firewood, and destruction of biodiversity (Maathai, *Unbowed* 123). As they have ceased to grow crops for consumption, they have been unable to feed themselves and have needed to buy their food, which is often processed and contributes to malnutrition (Maathai, *Challenge* 234–35). Their livelihoods have become more vulnerable to the vicissitudes of both weather and the global economy. Industrial agriculture, particularly on the kind of plantations described by Ocol, has played no small part in vastly uneven wealth distribution and conflict (over land and resources), as well as the impoverishment and disempowerment of the rural poor (Perfecto, Vandermeer, and Wright).

More generally, Okot's "songs" can be read as precursors to Maathai's narrative regarding the means of addressing the African nation's and Africa's ills. Like Lawino, she represents hope as rooted in traditional communal identity and culture formed by a close relationship with the soil, embodying an ideal ecological wisdom and enabling effective environmental practice. For example, she remembers her mother's warnings not to disturb wild fig trees, which as sources of streams and centers of biodiversity were protected by communal regulations tied to animistic belief: "When my mother told me to go and fetch firewood, she would warn me, 'Don't pick any dry wood out of the fig tree, or even around it.' 'Why?' I would ask. 'Because that's a tree of God.... We don't use it. We don't cut it. We don't burn it'" (*Unbowed* 44–45). When she is older, Maathai recognizes the wisdom of their practice: "I later learned that there was a connection between the fig tree's root system and the underground water reservoirs.... The reverence the community had for the fig tree helped preserve the stream.... The trees also held the soil together, reducing erosion and landslides. In such ways, without conscious or deliberate effort, these cultural and spiritual practices contributed to the conservation of biodiversity" (*Unbowed* 46). In *The Challenge for Africa,* Maathai extends this representation of an ecologically wise culture to the continent as a whole: "People . . . lived in harmony with the other species and the natural environment, and they protected the world" (161–62).

Many of Africa's current problems, according to Maathai, stem from the denigration of these cultures by colonialism and the spread of a capitalist outlook encouraging a narrow instrumentalization of nature and its rapid transformation for the sake of profit and economic development: "when Kenya was colonized and we encountered Europeans . . . we converted our values into a cash economy like theirs. Everything was now perceived as having a monetary value. As we were to learn, if you can sell it, you can forget about protecting it" (175). In the long run, the shift in cultural attitudes encouraging Ocol to reject indigenous food and agriculture and to focus on the pursuit of monocrop agriculture has led to disasters; the demonizing of African "indigenous cultures" led "to the virtual disappearance of the cultivation of many indigenous foods like millet, sorghum, arrowroots, yams, and green vegetables, as well as the decimation of wildlife, all in favor of a small variety of cash crops. . . . The loss of indigenous plants and the methods to grow them has contributed not only to food insecurity but also to malnutrition, hunger, and a reduction of local biological diversity" (Maathai, *Challenge* 175). While Maathai emphasizes the loss of both ecological and human health more than does Okot, both represent the erosion of indigenous culture as the root cause of a contemporary degraded landscape.

In turn, both also suggest that a cure for the ills bred by colonialism lies in a cultural return that will enable reclamation of self and a reconnection with the land. Throughout her song, Lawino tries to liberate Ocol from the control exerted by the effects of colonial education by reeducating him in the ways of Acoli culture and emphasizing their benefits; at the end of *Song of Lawino,* in her final appeal, she asks him to use the curative effects of traditional foods and medicines in order to regain his senses (literally), return to himself, and lead his people to a better future. Similarly, for Maathai, cultural reeducation is the "missing link in Africa's development" (*Unbowed* 175). Writing of her own growth, she claims "the tenets of modernity" to be "insufficient to provide an ethical direction for our lives": "Ultimately, I began to accept, and even yearn for, the part of me that had been concealed for so long, the part found in the culture into which I was born" (*Challenge* 162).

Maathai's alternative model of development necessitates a return to a natural guiding spirit embedded in the land: "For all human beings, wherever we were born or grew up, the environment fostered our values, nurtured our bodies, and developed our religions. It defined who we are and how we see ourselves" (*Challenge* 177). Alienation from their natural cultural identities "explained to [her] why many Africans . . . facilitated the exploitation of their countries and peoples. Without culture, they'd lost their knowledge of who they were and what their destiny should be" (*Challenge* 166–67). She acknowledges that precolonial cultures "had problems," especially in terms of social stratification and gender, but she still upholds the notion of ecologically ideal (and unitary) African cultures and of a fall from Eden when Africans "succumbed, not to the god of love and compassion they knew, but to the gods of commercialism, materialism, and individualism" (*Challenge* 165).

Finally, Maathai and Okot are linked by a sensibility we might associate with the environmentalism of the poor. Both point to the interdependence among precolonial agro-ecosystems, indigenous biodiversity, and the well-being of human communities, and both give the natural world value well beyond that of a narrow economic (monetary) instrumentalism. Yet they also remain focused on issues of resource control and utilization in terms of the well-being of local communities, the economically and politically disadvantaged, and future generations. In addition, they share a perspective on the intersection of gender and environmental concern associated with the environmentalism of the poor, in which women have played central roles both as leaders and as participants. This centrality is often explained in terms of women's positions in rural communities as those working most closely with natural resources and most attuned to the impact of environmental change; they are the community members who most often "gather fuelwood, collect water, and harvest edible plants" (Guha 108). In Africa, they represent a large percentage of small-scale farmers, and "the poorest of the poor are typically rural women" (Watts, "Visions" 125). In this sense, it is not surprising that the Green Belt Movement was focused, especially initially, as much on gender as on the environment: "as forests and watersheds became degraded, it was the women who had to walk the extra miles to fetch water and firewood, it was the women who had to plough and plant in once rich but now denuded land" (Nixon, *Slow* 140). Their work has also, of course, given women effective environmental knowledge and practice, as Maathai suggests through the figure of her mother. In

this context, Okot's use of Lawino to be the voice articulating the benefits of traditional environmental knowledge and agro-ecosystems links
his poems with Maathai's writing, as does the suggestion that Lawino
will be the one most negatively impacted by the changes wrought by
Ocol's form of development.

However, Okot's songs also reinforce a gendered national imaginary
against which Maathai struggled. As described by Anne McClintock,
this imaginary constructs women "as the symbolic bearers of the nation" but denies them "direct relation to national agency" by constructing their political roles in terms of familial relationships with men (354).
This "uneven gendering of the national citizen" is reinforced by the yok-
ing of gender to a split in national time (358). The modern nation looks
nostalgically to an archaic past embodying a timeless national essence
and forward to national modernity. This contradiction is managed
through the designation of men as the agents of national progress and
linear time and the alignment of women with the "atavistic and authentic body of national tradition" and "cyclical time" (McClintock 359, 377).
This temporal narrative can be understood in terms of the pastoral plot.
Women represent a component of the nation outside the public, historical sphere to which men come for inspiration and that they must protect
as they forge the nation's history. Part of this protection entails making
sure that women, as the vessels of national culture, are not corrupted
and do not go astray.

In the shadow of colonialism, this kind of narrative became a symbolic means for colonized men "to represent, create or recover a culture
and a selfhood that [had] been systematically repressed and eroded"
(Loomba 182). Struggling against a discourse that represented the colonizer as the proper patriarch reforming degenerate traditional family
relations and saving colonized women, anticolonial nationalists portrayed themselves as able to represent and defend the purity of tradition
embodied in women and the family and threatened by the pollution of
European modernity. As numerous critics have noted, colonized women
became the terrain over which colonialists and anticolonial nationalists
battled, and in the process, these women's voices and their perspectives
were suppressed.[1]

Okot's songs clearly reinforce the kind of gendered national narrative prominent in "the African literature of independence," in which
"men are invoked as leaders and citizens of the new nation" and "women

are . . . regarded as icons of national values, or idealized custodians of tradition" (Boehmer 224–25; Stratton). Okot not only constructs Lawino as the voice of traditional culture and embodiment of the land but also suggests that any gendered conflict has arisen as a result of colonialism. She is angry at being abandoned by Ocol and desires only the return of her power as "first wife" and "mother of [his] first-born" (120). She urges him to restore his "manhood" and take on his proper role of "prince" by embracing his true cultural identity (119, 116). At the end of her song, she has "only one request": that he let her "praise" him through traditional song and dance and show him the "wealth in [his] house" (120). In this sense, she embraces the mediated national subjectivity granted to her by patriarchal nationalism. Part of her role as vessel for tradition includes her rejection of Western education, which can be linked with the fear of the overeducated and decultured woman who will challenge traditional male power.

This kind of anticolonial nationalist discourse, often articulated in the language of authenticity, continues to be used to curtail women's agency and prevent them from challenging restrictive gender roles and patriarchal power. In *Unbowed*, Maathai narrates her own familiarity with this situation. The first woman in East Africa to get her PhD, she took an active leadership role in national politics and the transformation of society and culture. Part of this transformation included giving women a new sense of agency and challenging the gendered configuration of the nation, the family, and the nation-as-family. Early on, she learned that men were "threatened by the high academic achievements of women like me" and was accused of being "too educated, too strong, too successful, too stubborn, and too hard to control" (*Unbowed* 139, 146). The male authorities were especially threatened by "an educated, independent African woman aspiring to leadership" and by her bringing "to the national arena the issue of power and gender" (157, 159). "In a country where a good African woman was not supposed to be involved in politics," she was represented as too Westernized, labeled "disobedient" by the Moi regime, and threatened with forcible circumcision by a member of Parliament (235, 159, 244). Moi himself "suggested that if [she] was to be a proper woman in 'the African tradition,'" she "should respect men and be quiet" (196). As she tells her story, she brings attention to how the type of gendered discourse of authenticity served up by Okot is used to squelch women's empowerment and, more generally, to suppress dissent against men in power who represent themselves as the defenders of tradition.

To argue for this distance between the two writers is not to claim that Maathai's own discourse escapes the political pitfalls of pastoral nativism. As noted earlier, she repeatedly offers images of an idealized ecological indigenous identity and makes calls for a return to homogenous cultures unsullied by social contradiction. In fact, her writing is fraught with the tension between discourses challenging and upholding notions of cultural authenticity and purity. This tension separates her from Okot, who mostly offers a monologic (gendered) nativism, but it does not do away entirely with the pitfalls of a narrative often focused on a colonial fall from ideal, authentic, natural identity. This notion of a fall locates the primary problem in the legacies of past colonialism and in culture; if Kenyans can just rid themselves of the outlook fostered by British rule, then they will be able to address the limitations of the postcolonial state and decolonize Kenya. Such a narrative risks eliding or deemphasizing current, global relationships, which might be described as reformulated forms of imperialism and which restrict or complicate the possibilities for change. In other words, the solution of return to a culture untransformed by the hybridizing effects of history risks simplifying the conditions that need to be understood if the problems facing Kenya are to be addressed in the long run. It also serves to make the Western reader comfortable, which may, in fact, partly explain its prominence in *Unbowed;* she or he can be urged to contribute to the Green Belt Movement but does not need to reflect deeply on how her or his material conditions of existence might play a fundamental role in the political, cultural, and ecological dynamics in Kenya described by Maathai.

As already indicated, Maathai's writing and Okot's songs also differ in terms of the priority given to environmental issues and of concomitant attention to ecological relationships. Maathai's emphasis on the importance of understanding the details of those relationships and on the necessity for Africans to make environmental degradation a central concern is clearly absent from *Song of Lawino* and *Song of Ocol.* To claim that Okot's songs, or Ngũgĩ wa Thiong'o's *A Grain of Wheat,* anticipate Maathai's writing and that they could be considered as part of a legacy of environmental writing in East Africa is not to say that they have the same degree of engagement with ecology and actual environmental change. Yet this difference in focus does not necessarily mean that her writing represents an unambiguous or uncomplicated progression—even in terms of environmental discourse.

Ngũgĩ wa Thiong'o's *A Grain of Wheat* highlights the themes of betrayal and disillusionment in the newly independent African nation, as well as the search for a transformed anticolonial struggle. While writing the novel, Ngũgĩ had become strongly influenced by Frantz Fanon, who famously declares in *The Wretched of the Earth* that the "battle against colonialism does not run straight away along the lines of nationalism" (Fanon 148)—that is, that the goal of national independence was not necessarily identical with the goal of dismantling an unequal, exploitative colonial system and culture because these could live on, albeit in new forms. The novel takes place at the moment of Kenya's independence in 1963, but the majority of the narrative focuses on the period between the end of World War II and the defeat of the Mau Mau liberation movement in the mid-1950s.

A Grain of Wheat begins with a depiction of the present as degraded and corrupt, with the hopes of a new future through independence already blighted. Environmental conditions connote this degeneration; in the first chapter, the land is described as "sick," "dull," "dry," and "hollow" (6). In the course of the novel, these conditions are associated with the corrosive alienation and isolation caused by colonialism, both among people and between people and land. The degraded present is contrasted with images that allude to a time when, as a result of indigenous Gikuyu cultural identity, communal bonds among people were strong and the land was healthy and beautiful. In this sense, an idealized past and the harm caused by colonialism are represented in fairly mannered pastoral terms. However, the narrative also offers a strong historical vision that focuses on the complicated transformations wrought by colonialism throughout Kenya and suggests the need for adaptation—for example, a transformation of the anticolonial struggle, rather than simple return. Particularly pertinent for an ecocritical reading of the novel are Ngũgĩ's depictions of the environmental impact of specific colonial attitudes, policies, and practices.

In *A Grain of Wheat,* the ideology of geographic difference and the desire for mastery underpinning colonialism are summed up in the title of a book the colonial official John Thompson seeks to write: *Prospero in Africa* (54). For Thompson, Shakespeare's imperial sorcerer-king invokes the image of the heroic colonial mission to tame and make useful savage wilderness. Instead of magic, however, the British use science; Thompson's final administrative post in Kenya is as codirector of the "Githima Forestry and Agricultural Research Station," created to be part of "a new colonial development plan" (33). Such stations were part of the project of scientific conservation found throughout the British Empire. The proclaimed goal of this work was the development of techniques and policies for sustained yield and the rational use of natural resources, resulting in the prevention of unsustainable environmental degradation, particularly from deforestation and soil erosion. However, as environmental historians have shown, conservation and environmental science were a means for colonial governments to exercise control over land, people, and resources.[2]

In the case of scientific forestry, the creation of forest reserves and powerful forestry departments enabled the state to seize vast areas of forest already in use by indigenous peoples who, it was claimed, did not know how to take care of the bounty given to them by nature. Furthermore, while such policy was apparently intended to prevent unrestrained exploitation of natural resources for short-term gain, it also worked closely with capitalist development, often with disastrous results; the goal was the establishment of an efficient system of resource use that allowed for the greatest profits for the longest time possible. In the case of agricultural science, the claim that traditional farming practices caused soil erosion became an excuse in Kenya to justify land alienation and resist redistribution (Mackenzie, *Land*). In addition, in Kenya "discourses of 'betterment' and 'environmentalism'" were used to deepen administrative control over African farmers (Mackenzie, "Contested" 699). They were pushed to embrace the practices of scientific agriculture in order to preserve soil and increase yield and to transform the Gikuyu Reserves into spaces that would resemble British rural landscapes (Mackenzie, *Land* 701). This transformation had an economic goal: an increase in yield and in marketable crops leading to an increase in tax revenue (703). In this sense, colonial environmental science and policy in Kenya encouraged the spread of capitalized agriculture, as well as the class divisions that came with it.

A Grain of Wheat represents the spread of a capitalist ethos among Gikuyu farmers as socially and environmentally destructive. Mugo, who betrays a leader of the resistance to the British, focuses on capital accumulation through his agricultural work, with the ultimate objective of social recognition: "He would labour, sweat, and through success and wealth, force society to recognize him. There was, for him, then, solace in the very act of breaking the soil" (8). In the course of the novel, Mugo's connection with the environment, based on the land's utility for achieving dreams of wealth and power, becomes associated with the kind of degraded landscape he struggles to cultivate in the opening of the book: "[Mugo] raised the jembe, let it fall into the soil; lifted it and again brought it down. The ground felt soft as if there were mole-tunnels immediately below the surface. He could hear the soil, dry and hollow, tumble down" (6).

Such descriptions of environmental desiccation seem primarily to serve as a means to figure the corrupt condition of the nation resulting from the inability to break with colonial capitalism (a reading in line with the growing influence of Marxism on Ngũgĩ). The focus seems to be not so much on actual environmental degradation and its causes, as on degenerate social relations characterized by isolation and betrayal, which environmental conditions represent. Yet the ecological signifi-cance of the connection *A Grain of Wheat* makes between the impact of colonialism and an infertile land should not be dismissed, especially in light of what we know about the changes wrought by the colonial environmental attitudes and policies Ngũgĩ highlights. The images of a dry, unproductive land suggest soil erosion and infertility, inadequate water supply, and poor crops—which were, and remain, long-term re-sults of the impact of scientific agriculture and conservation in Kenya. In the reserves, new crops and new crop varieties, in addition to being less drought and disease resistant, "together with new agricultural prac-tices such as pure cropping, clean weeding and deep tillage, contributed to the acceleration in loss of soil fertility" (Mackenzie, *Land* 702). The deforestation accompanying cash crop farming also contributed to soil erosion, as did the kind of colonial forestry associated with places like the Githima Research Station. Lacking an understanding of the benefits of tropical biodiversity and focused on the production of commercially desirable woods, the British fostered the growth of monocrop, nonindig-enous tree plantations in Kenya; these plantations have contributed sig-nificantly to the loss of soil, watersheds, and rivers (Maathai, *Unbowed* 38–39, 121–22; Maathai, *Challenge* 244–45). Thus, the link between the legacies of colonialism and environmental degradation alluded to by Ngũgĩ might be considered historically and ecologically astute rather than simply a recycling of pastoral tropes that downplay issues of real environmental threat.

In *A Grain of Wheat*, the struggle against the legacies of colonialism necessitates a rejuvenated anticolonial movement. Ngũgĩ suggests it will need to be different from previous manifestations of the movement, since what must change is not foreign, white rule but the structures and psychology it set in place. In this situation the primary task of an anticolonial movement is to foster self-sacrificial honest dialogue and confession, revealing the extent to which the nation and its citizens have become internally colonized. Only through such dialogue and confession will the new nation overcome the atomization, focus on self-interest, exploitation of other people and the land, and betrayal of the hopes for a better future for all Kenyans. Yet if the movement must be transformed, it will also need to remain closely tied in spirit and principle to the anticolonial struggle of the past. In particular, it must draw from that struggle a mode of understanding the world very different from the one offered by colonialism and capitalism, one based on communal interdependence and responsibility, as well as on a corresponding willingness for self-sacrifice (Caminero-Santangelo 49–67).

Finally, it needs to pursue the "bond with the soil" from which anticolonialism's "main strength" always "sprang" (Ngũgĩ, *Grain* 12). This bond is based on familial love and identification rather than on private ownership. As the resistance leader Kihika proclaims, "This soil belongs to Kenyan people. Nobody has the right to sell or buy it. It is our mother and we her children" (98). Ngũgĩ links this ethic with Gikuyu culture; in the novel, indigenous dwelling is defined by the interconnection between human habitation and a local nature that has not been repressed through the effort to master it. In the days before the rebellion, "a bush—a dense mass of creepers, brambles, thorn trees, nettles and other stinging plants—formed a natural hedge around the home" of a "a well known elder in the ridge." These hedges were part of all the homes in the village of Thabai and "were hardly ever trimmed; wild animals used to make their lairs there" (75). The connection between the liberation struggle and Gikuyu environmental attitudes and practice is perhaps best exemplified through the use of the forest during the uprising. If for the colonists, the forest represents all that must be transformed if Africa is to be rendered useful and livable, for the resistance fighters it is a place of refuge and protection in which they live and organize; historically, it was a kind of home for the Mau Mau movement (Nicholls 184–86).

The kind of relationship with the land that Ngũgĩ associates with anticolonial struggle and with Gikuyu culture offers the possibility of a different conception of development from the one pushed by colonial modernity. The traitor Karanja betrays the resistance because of his desire for "whiteman's power . . . a power that had built the bomb and transformed a country from wild bush and forests into modern cities, with tarmac highways, motor vehicles . . . railways, trains, aeroplanes and buildings whose towers scraped the sky" (156). In *A Grain of Wheat,* those like Karanja, as they become leaders, betray the land and the majority of the people. A revitalized and healthy new nation requires a model of development based less on technological and economic growth and on mastery and more on an ethics of care and responsibility rooted in reestablished bonds within communities and between people and the land. The distance between *A Grain of Wheat* and Ngũgĩ's later novel *Petals of Blood* (1977), which also offers "an indictment of the environmental damage produced by colonialism and neocolonialism" (L. Wright 32), is a relative deemphasis in the latter on more locally generated solutions and on return to previous relationships among land and people as a source of hope in favor of a more doctrinaire Marxist teleological narrative.

Maathai's exploration of independent Kenya's colonial inheritance and of the need for a liberatory movement based on principles from traditional Gikuyu culture has substantial overlap with Ngũgĩ's. She too suggests that Kenya's ills have stemmed from notions of development based on individual acquisition of wealth and power, on strictly technical and industrial advances, and on GNP (as opposed, for example, to distribution of wealth). Like Ngũgĩ, she urges a shift from narrow individual or group interests to a larger sense of communal interest, from a conception of development that benefits only a few to one focused on the well-being of all Kenyans and on social equity. Also like him, she encourages a kind of self-sacrifice. "Africans," she claims, "must begin the revolution in ethics that puts community before individualism, public good before private greed, and commitment to service before cynicism and despair" (*Challenge* 23). This ethic, which underpinned the Green Belt Movement, was revolutionary in the context of Moi's regime: "Against the backdrop of Kenya's winner-takes-all-and-takes-it-now kleptocracy, the movement affirmed a radically subversive ethic—an ethic of selflessness" (Nixon, *Slow* 135). Part of the decentering of the self and its interests involves a shift not only in communal obligations but also in temporal relationships: "I have to keep reminding [people] that the trees they are cutting today were not planted by them, but by those who came before. So they must plant the trees that will benefit communities in the future" (Maathai, *Unbowed* 289).

As was the case with Okot, Ngũgĩ does not place nearly the same emphasis on the centrality of ecological awareness for a better future as does Maathai. *A Grain of Wheat* suggests that those who wage an anticolonial struggle automatically speak and fight for nature's interests. In this sense, there is no need to give ecological understanding any primacy; that understanding will come with the rejection of colonial epistemology. In contrast, Maathai emphasizes the need to acquire ecological knowledge if the future is to be secured against the threat of what Nixon terms the "slow violence" of environmental degradation: "When a river dries up in Kenya or a crop fails, people tend to pray to God for more rain and food. . . . What they don't do as much as they should is ask *why* the river has dried up or crops have failed—questions that involve a deeper analysis and a more holistic approach to the management of ecosystems" (Maathai, *Challenge* 248). In Maathai's writing, the fight against colonial values is crucial in the struggle for environmental health, but it is not adequate. Compared with Maathai, Ngũgĩ's romantic pastoral downplays actual environmental degradation, its causes and effects. In addition, he, like Okot, embraces a patriarchal nationalist discourse that feminizes the land (Nicholls)—for example, when Kihika represents the land as "our mother"—with many of the same problematic implications for women's agency and for environmental movements organized by women.

Nevertheless, it might be too easy to represent Maathai's writing as offering a clearly more advanced environmentalist sensibility than *A Grain of Wheat* and, concomitantly, to gesture toward a linear, progressive narrative for environmental writing in East Africa. For example, she reinforces an ethos of the autonomous individual more than does Ngũgĩ, even in his early writing in *A Grain of Wheat*. As Nixon notes, *Unbowed* is a memoir in which "collective history" is "recast . . . as a personal journey with a singular autobiographical self as its gravitational center" (*Slow* 142). This form may be, as Nixon suggests, in large part due to the pressures of Western publishers and audiences, as well as the genre of the struggle memoir, but it still contrasts sharply with the narrative form of *A Grain of Wheat*. The narrator of Ngũgĩ's novel is an unnamed member of a Gikuyu village who speaks as an equal to other members of that village and references shared memories, history, geography, and communal identity. In this sense, there is less focus on individual achievement and importance than in Maathai's memoir (or in *The Challenge for Africa*). She, more than Ngũgĩ, reinforces the image of the heroic autonomous self, a notion that can be problematic for those encouraging notions of ecosystemic interdependence and a transformed conception of individual identity in the fight against environmental change.

The authority of this heroic self is naturalized in part through Maathai's appeal to anticolonial pastoral ideals. She represents herself as having been able to step away from the false, destructive values of colonial modernity and to connect fully with the authentic spirit of the land and people before most Kenyans. As a result, her authority to guide them and represent their interests was only natural. In a reversal of patriarchal national discourse, she replaces the father of the postcolonial nation with a mother; however, the connections between her autobiography and the underlying topology of that discourse remain striking. In his *Blood in the Sun* trilogy (*Maps, Gifts, Secrets*), Nuruddin Farah undermines precisely the conceptions of identity, knowledge, and authority encompassed by such a topology.

Pastoral and Its Discontents: Camara Laye, Nuruddin Farah, and Wangari Maathai

In their emphasis on the agency afforded by reclamation of natural selves and cultures, Wangari Maathai and Okot p'Bitek evoke the anticolonial pastoral trope of empowering, magical secrets famously deployed by Camara Laye in *L'enfant noir* (*The Dark Child*) (1954). This autobiographical tale of childhood in the highlands of Guinea ends with Camara's journey to France as a young man in pursuit of his education; alienated and nostalgic, the narrator remembers his past in the most idealizing terms. He depicts Malinke culture as enabling connections with the secrets of the self and local nature and as affording an effective, authentic authority. Such a culture, the autobiography suggests, results in a healthy, peaceful, and just society in alignment with the natural world.

This idealizing representation is established early on in the narrative through the description of Camara's father's totem and its significance. Established by patrilineal ancestry, the totem embodies a deep, natural identity defined in biological and religious terms; blood becomes a manifestation of transcendent spirit. Camara's father's totem is a small black snake, and his connection with it allows him access to important knowledge of the natural world and of the future. In turn, it grounds his extraordinary skills as a smith and his authority within the community: "My name is on everyone's tongue, and it is I who have authority over all the blacksmiths in the five cantons. If these things are so, it is by virtue of this snake alone. . . . It is to this snake that I owe everything; it is he who gives me warning of all that is to happen" (25). Camara's father understands nature's secret language, an understanding that bequeaths mastery; while working with gold, his "lips moved and those inaudible, secret words, those incantations he addressed to one we could not see or hear, was the essential part. Calling on the genies of fire, of wind, of gold and exorcising the evil spirits—this was a knowledge he alone possessed" (36). To maintain this connection with his "guiding spirit," he must follow the proper behavior and rituals as dictated by Malinke culture (26). For example, before working in gold, he would enter his workshop "in a state of purity, his body smeared with the secret potions hidden in his numerous pots of magical substances" (37). These potions reflect a traditional knowledge of local nature; they are made with "extracts from plants and the bark of trees," and each "had its own particular property" (19). In addition to giving access to effective mastery of the material world, traditional Malinke culture's close relationship with local nature seems to foster impeccable virtue; the father's sense of justice is foundational for the health and harmony of both family and community. Thus, patriarchal authority is legitimated by being made to appear *only natural.*

Camara subtly contrasts the morality and naturalness of rural life among the Malinke with the corruption of colonial modernity. In particular, colonial education is associated with the acquisition of abstract knowledge and forms of bureaucratic authority separated from the underlying spirit—the secrets—of nature and community. This separation, Camara suggests, results in an increase in injustice, the degeneration of well-being, and conflict in the community. The schools Camara attends are unhealthy environments rife with cruelty, inequality, arbitrary authority, and the threat of infection and/or starvation. More generally, the narrator often seems to wonder if he and, to a lesser extent, his community have regressed rather than progressed under colonial tutelage. The decrease in reverence for and connection with a local natural environment results in the loss of useful and powerful knowledge (Camara asks at one point, "Do we still have secrets?" [109]), as well as of healthy relationships with natural and human communities. Meanwhile, colonial education takes Camara further and further away from an unalienated, natural identity. The bridge to a rooted, authentic self, embodied for Camara's father in his relationship with his totem, eludes the son. Noting the "appealing caress and the answering tremor" that constitutes a kind of conversation between man and snake, Camara laments that he would have liked "to understand and listen to that tremor too," but he "did not know whether the snake would have accepted my hand" or have anything to tell him: "I was afraid he would never have anything to tell me" (28–29).

Yet, despite its anticolonial elements, *L'enfant noir* is the least counterhegemonic of the pastoral narratives discussed here. Christopher Miller offers an eloquent defense of Camara against the many accusations by "African critics" of "apolitical" quietism, but the reader being "seduced into a village idyll" remains all too likely (Miller 122–24; Soyinka, "From" 387). Colonial relationships remain mostly off stage, and the narrative often seems to suggest that Camara's alienation and the erosion of Malinke culture result from inevitable, universal processes. To a significant degree, nature and tradition are represented as pre- or ahistorical while the modern is the location of change and history. Furthermore, the mother figure is associated strictly with stasis and atavistic time, but the father remains both connected with tradition and able to look ahead. (He pushes the narrator to pursue his colonial education and to go to France.) Thus, Camara serves up notions of authenticity and authority associated with patriarchal nationalism (discussed in the previous section of this chapter) that have proven highly problematic in postcolonial Africa.

In *Secrets,* Nuruddin Farah offers a counterdiscourse to the patriarchal pastoral narrative of *L'enfant noir;* in particular he interrogates the relationship between this kind of narrative and disastrous models of identity in the postcolonial nation. Rather than enabling access to an authentic self and harmonious connection with the natural world, authoritative secrets of the sort depicted by Camara are, in Farah's novel, objectifying, alienating constructs closely tied to exploitative forms of power. They do not unveil natural authority but naturalize socially constituted inequality and injustice. Farah is also concerned with how the intertwined, hierarchical sets of terms structuring many anticolonial pastoral narratives (authentic/inauthentic, natural/unnatural, pure/impure) enable the shifting of responsibility for Somalia's problems. As he complains elsewhere, "Somalis do not place themselves, as individuals, in the geography of the collective collapse, but outside of it" (*Yesterday* 187–88).

In contrast with Camara, Farah repeatedly undermines efforts to locate the self and others using a vocabulary of animal totems. Such efforts are complicated by the multiplicity of animals associated with one person and by the plethora of often conflicting interpretations of the meaning of animals. Such complications plague the effort to use a crow, present at the protagonist Kalaman's birth, as a natural key to his identity. For one thing, the meaning of the crow is muddied by differing interpretations resulting from varied cultural vocabularies. For another, the crow is not the only winged creature associated with his birth—there is a sparrow and a bee—and more generally, multiple animals have contributed to the development of his personality. Says Nonno to his grandson, Kalaman, "Many other creatures had preceded the crow's arrival, many more came after his departure. I can think of an ostrich or two . . . mail pigeons, a peacock, ducks, a skyful of birds *and* Hanu [a pet monkey]" (101–2). Finally, the connection between humans and animals is used in fundamentally contradictory ways. On the one hand, drawing on a standard human/animal binary, characters compare others to animals as a means of denigration; on the other hand, those identifying themselves by clan use totemic animals that can supposedly define group filiation and the powers with which that filiation is associated. Such a contradiction points to the idea that the conceptions of nature embraced by the characters are part of a socially constructed discourse.

At the same time, *Secrets* reflects a concern with the nonhuman natural world in ways that had not been apparent in Farah's previous writing. The novel emphasizes that if the exploitation and instrumentalization of women are linked with their alignment with nature, this construction also helps position nonhuman "others" as only having value to the degree to which they can be manipulated and used to satiate desire. In this sense, his novel can be associated with an ecofeminism concerned with the interconnections among various dualisms (man/woman, culture/nature, subject/object, reason/emotion, male/female) and how these interconnections tautologically contribute to both the oppression of women and the exploitation of nonhuman nature.[3] (For example, women's objectification is legitimated through their association with nature, while nature, by being feminized, is positioned as in need of male mastery.) In *Secrets,* women are frequently denigrated or ostracized through the use of animal analogies—mostly, but not exclusively, by men. Sholoongo, who since birth has been constructed as an outsider, is frequently associated with dangerous or disgusting animals. In a conversation with Sholoongo's half-brother, Kalaman identifies her as a rat, and her brother characterizes her as a maggot. However, Kalaman recognizes the gendered implications of these associations: "Here we were, two men, one half a brother to her, the other once infatuated with her, mean men bad-mouthing a woman whom they called a bitch, witch, a whore" (55). The novel links the denigration of nature implied in women/animal analogies with actual environmental destruction in Somalia in the last decades of the twentieth century. Farah focuses, in particular, on the killing off of Somalia's wildlife and on deforestation for the sake of individual profit through international trade. During the unrest of the 1980s and continuing throughout the 1990s, corrupt government officials, businessmen, and clan warlords, taking advantage of growing anarchy and the disintegration of the nation, cut the forests and sold the wood for charcoal to Middle Eastern businessmen while selling off prized pieces of protected animals to Asian middle men (Lacey). In turn, these unreported or, at least, underreported environmental catastrophes contributed to ever-growing loss of livelihoods for farmers, famine, civil war, and continued manipulation of Somalia's affairs by numerous foreign nations and groups.

The ecofeminist elements of *Secrets* clearly link the novel with Maathai's narratives. Maathai too is focused on the connections between women's rights and the struggle against environmental degradation in the postcolonial nation, and she fought against notions of authenticity used to legitimate oppressive gendered relationships and the plundering of the nation's ecosystems. Yet Farah differs from Maathai in terms of the degree to which he undermines the underlying binary of the authentic and inauthentic. She holds on to the idea of a secret key to the true self and, more generally, of authenticating secrets embedded in geographical place. Farah, in contrast, remains unrelenting in his anti-pastoral focus; he suggests that appeals to authentic identities and for return to forms of knowledge and authority rooted in a perfect harmony with local nature all too easily lend themselves to exclusive and centered sociopolitical models, to the curtailment of ideas of responsibility, and consequently to exploitation and degradation. For Farah, truly

counterhegemonic models of collectivity must foreground histories of
mutual transformation and a recognition that all relationships are political and ethical rather than naturally or spiritually given. They must
also include every kind of social institution. In particular, for Farah, the
transformation of the family is a metaphor for and a necessary part of
positive change in the reconstitution of the nation.[4]

Reading *Secrets* in relation to Maathai's activist writing further
highlights the potential limitations in her narrative of identity discussed
in the previous section. At the same time, however, such contrapuntal
reading also points to potential pitfalls in Farah's skewering of pastoral
tropes and in his focus on the decentering of consciousness as a locus
for the reimaging of resistant subjectivity. Maathai's successes as an activist were not necessarily achieved *despite* her deployment of pastoral
tropes but *through* them, while Farah's more postmodern perspective
potentially suppresses structural issues and might even hamper the mobilization of resistant identity.

The final installment of Farah's *Blood in the Sun* trilogy, *Secrets* takes
place in the days preceding the full onset of Somalia's collapse into
clan warfare in 1991 after the dictator Siad Barre fled into exile. Not
surprisingly, an obvious concern in *Secrets* is "the damage that clan
fanaticism . . . has done to [Farah's] country" (D. Wright 125). The novel
emphasizes that such fanaticism is not determined by atavistic, precolonial tradition but by political interests shaped by imperial modernity and manipulating the concept of natural, sacred allegiances. As Michael Eldridge notes, in *Secrets* "Somalia's chronic clan-riddeness (like
Africa's 'anarchy') is a condition that's not congenital, the product of
begats, but political, the product of maneuvers and machinations" (653).
This point is articulated by Farah's protagonist Kalaman and his grandfather; they suggest that the colonists encouraged Somalis to identify
themselves by clan in order to facilitate control and that clan-thinking
is still determined by political ambitions:

> What makes one kill another because [his/her] mythical ances-
> tor is different from one's own: this has little to do with blood,
> more with a history of the perversion of justice. What makes one
> refuse to intermarry with a given community has to do with the
> politics of inclusion and exclusion. Forming a political allegiance
> with people just because their *begats* are identical to one's own—

judging from the way in which clan-based militia groupings were arming themselves—is as foolish as trusting one's blood brother. (76–77)

According to Kalaman, the concept of clan is based on a confusion of biological and political identity, and the notion of natural collectives suppresses awareness of differences in interests and power within communities. Such naturalization contributes to perverse, unjust killing and marginalization of "others" and, ultimately, results in foolish self-destruction. In a sense, the novel sanctions Kalaman's summation; in *Secrets*, a focus on defining origins located before (or outside) history and politics is a significant factor in Somalia's collapse. Yet Kalaman and Nonno's attribution of blame to clan thinking also suppresses the role patriarchal anticolonial nationalism and their own adherence to its pastoral narratives play in the catastrophe of the postcolonial nation. As is true throughout the *Blood in the Sun* trilogy, *Secrets* suggests this nationalist discourse too naturalizes social identity and authority in pernicious ways.

Nonno is the patriarch of his family and farm, and his authority is anchored by a reputation for a magical knowledge and mastery of the natural world. He can apparently speak the language of animals and control the inanimate forces of nature. He also seems to have a deep understanding of the defining characteristics and motivation of other people. He is frequently compared with "King Soloman," who had "a revered standing among the practitioners of the science of magic" and "whom Allah helped 'subject the wind blowing strongly'... Soloman upon whom Allah 'bestowed knowledge in judging men' and whom Allah taught to speak 'in the language of birds and other matters'" (239–40). As a result of this reputation, many characters believe he can effectively manage and protect his lands, the people who live on them, and his family. Nonno himself understands knowledge in terms of an accumulation of defining secrets: "they mark us, they set us apart from all the others. The secrets which we preserve provide a key to who we are, deep down" (144). He is obsessed with gathering them and believes they give him the ability to shape and control his world. Like Camara's father, he embraces the image of the traditional patriarch with powers of "articulation" stemming from mastery of nature's language (275); at Kalaman's birth, Nonno "forged" into his daughter-in-law's room, "his lips ashiver with the totemic powers of the crow, to whom we prayed long ago as our skygod, the crow who was revered as a deity among the peoples of the Horn of Africa before the spread of Islam and Christianity" (162).

As this scene suggests, Nonno's access to authoritative secrets seems to be enabled (especially from his and Kalaman's points of view) by his adherence to precolonial tradition and his freedom from colonial ideology. In turn, the forms of social order associated with his patriarchal authority are legitimated by the suggestion that they transcend colonialism's false consciousness and are grounded in an unalienated relationship with the land. When Nonno was younger, Kalaman tells us, he had "once refused to have the name of his clan in his identity card, as was the custom in the Italian colony," and as a result, "he spent a while in detention, accused by the Italians of being an anarchist" (193). The connection made by the colonists between anarchism and a refusal to be identified by clan points to the relationships among a colonial epistemological order, determinate (essentialized) identities, and political control, and the story itself suggests that Nonno has recognized this relationship and refused the loss of agency that would come with an acceptance of it. For him, liberation accompanies the formation of a national identity apart from clan: "One's 'Somaliness,' as opposed to being identified as belonging to a given clan, defines a political entity. The clan is nonpolitical, based as it is on one's primordial blood identity" (193). Yet, while Nonno rejects the notion of clan because it elides politics, he still constructs his own patriarchal authority as natural and "nonpolitical." If in colonial discourse authority was legitimated by mastery of nature through science, an anticolonial nationalist narrative grounds patriarchal authority in knowledge of and identification with nature through adherence to tradition. In Nonno's model of postcolonial authority, the "natural" Somalian patriarch replaces the colonial overlord; in place of the British Prospero (from *A Grain of Wheat*), there is an African sorcerer-king.

Kalaman's own subjection to patriarchal, nationalist discourse has profound implications for his relationship with his grandfather. Nonno has enormous control over him, in part because Kalaman sees Nonno as offering an escape into an Edenic landscape away from the horrors and confusion of life in Mogadishu. For Kalaman, Nonno "was more of a place than a person": "What a pleasure it was for me, as a child, to know the geography of his person. What a delight it was for me to touch base with the symmetry of his vision, through contacts I made with his mind. I . . . sat on his lap as he gazed at the stars, repeating to me myths of old" (94–95). Understood spatially, Nonno's "place" is his rural "estate" ("Afgoi meant Nonno" [94]), embodying an apparent pastoral harmony between humans and nature and organized on natural principles: "It was often said that when in Nonno's bungalow, the seasons, like a tamed tigress, fed off the palm of your hand" (98). Kalaman even indicates that Nonno's knowledge of Somali culture has given him not only a profound grasp of "the interdependence of the human and animal worlds" but also the ability "by studying gazelle's mating-time" to "decide as to the auspicious nature of the seasons" (98). Throughout the novel, this image of Nonno as a sage able to foretell the future and shape events through his close relationship with nature gives the grandfather incredible power over his grandson.

Kalaman's mother, Dumac, worries about this power, and Kalaman unwittingly also suggests that his grandfather's apparently benevolent authority should not be taken at face value. She sees Nonno's pursuit of secrets as threatening: "The man has a penchant for secret-nurturing, secret-feeding, and secret-finding. He is not at all embarrassed to search for secrets among the Kleenexes in a woman's handbag" (149). Dumac's framing of Nonno's secret gathering puts a malignant spin on his reputation as a kind of Casanova. If Kalaman suggests that women's attraction to his grandfather comes from his size and from his "charisma," she believes he manipulates them using the knowledge he gets from an unseemly invasion of their privacy. Furthermore, Kalaman's own representations of Nonno can suggest he may not be a benevolent, protective patriarch. When a crocodile attacks "one of his laborers," Nonno sends his factotum Fidow to kill it. Yet "common knowledge" recognized that this "particular crocodile" was a "greedy beast" that had "made off in the recent past with two little girls and their mother" (9). It seems that Nonno only acted when his economic arrangements were threatened. Such moments in *Secrets* bring into question a gendered nationalist pastoral narrative representing the native patriarch as the voice and protector of women's and children's interests. Sholoongo perhaps best sums up the novel's skepticism regarding the benevolence of Nonno and his form of authority. She tells him that his claim to gather and keep secrets "for the general good of society" is a pretense and suggests that, in actuality, he perpetuates societal injustices: "You know what's wrong with our people? Where there is no individual justice, there can be no communal justice, certainly no possibility of democracy" (280).

There is also some question as to whether Nonno is actually taking care of the natural world of the estate or just exploiting it. If the killing of the "particular crocodile" would seem necessary, Fidow's slaughter of crocodiles on the estate in order to sell their skins is only about their extermination for financial gain (from which Nonno also benefits). Nonno himself has a huge menagerie of animals on the farm, with whom he often interacts, but their primary purpose seems to be as circus animals affording him the opportunity to display his power. Rather than seeing him acting as the protector of animals, we observe him instrumentalizing them for his own uses. Thus, he has trained "several species" of pigeons "as message carriers": "a million times more trustworthy than the nation's mail services, which were nonexistent, and much cheaper and more discreet than humans bearing verbal messages" (70).

The relationship between Nonno's form of authority and the exploitation of humans and animals is even more obvious in the representations of other patriarchal characters. In order to slaughter the crocodiles, Fidow attracts and renders them docile by learning and using their secrets—the smell they "emit just before mating" and the bellow a bull emits when preparing to mate (60). The links among the use of "secrets," sexual exploitation, and the instrumentalizing of the natural world is also made apparent through the character of Sholoongo's father. Even more than Nonno's, Madoobe's knowledge of animals seems to give him control over them, cultural authority, and wealth. He makes "his living taming wild horses, which he exported to the Middle East and out of which, it was rumored, he made a mint. What's more he was reputed to have been the first Somali ever to employ an ostrich as a guard to mind his horses and zebras, a feat which turned him into something of a celebrity" (10). Also like Nonno, Madoobe is a great seducer of women. Interestingly, when Kalaman asks Sholoongo about a scene he witnessed in which her father mates with a cow and seems to render it docile by speaking in "a secret language," she claims that this is a symbolic ritual. "'It was a cow,' Sholoongo said, 'whom my father has decided to domesticate, that's to say, take as his wife.' A couple of days later, Madoobe brought home a young bride" (17). Both women and animals are positioned here as objects for male, human gratification that are mastered by gaining access to their defining secrets.

The kind of patriarchal pastoral narratives embraced by Nonno and Kalaman are also undermined by the mistakes characters make when they believe they know defining secrets and, more generally, by the suggestion that belief in such knowledge fails to account for the evasive-ness of identity and leads to a dangerous lack of humility. Particularly striking is a moment when Nonno misreads the condition of a visiting plover by applying notions of nature separated from the constraints of culture. When the plover perches on a table at which Nonno and Kala-man are talking, Nonno challenges a cultural construction of the bird as ill omened and offers what he sees as its true significance: "Our visitor is no domesticated Hanu [their pet monkey] whom we pamper with a per-sonal name and our affection. This . . . plover is a freeborn, freethinking bird" (155). Carefully observing the plover, Kalaman contradicts Non-no's effort to turn it into the image of a pastoral nature free from human restraint. He notes to himself that there is a "thread . . . tied to the bird's shank" with "a tiny piece of paper": "Was the plover a message-carrying bird in our dining room? Contrary to Nonno's view, the bird was no free agent, roaming the winds at will" (155). Concomitantly, when Nonno tries to demonstrate his skills "at uttering bird noises" the results un-dermine his reputation for mastery in animal communication. When he makes his bird calls, not only do "all manner of birds" show up, but so do "ants," "the odd dog," "a giraffe," "a stallion," and "Hanu" the do-mesticated monkey. When he then tries to make an "appeal" in "pigeon parlance," most of the birds fly away, but those that stay include "crows, vultures, hawks, and a canary of sterling beauty" (102).

More seriously, Fidow's belief in his ability to understand the animal world and exploit it with impunity leads to his own destruction. After he kills off half a herd of elephants in order to sell its ivory to a group of "Kenyan, Somali, and Hong Kong businessmen" (101), an elephant comes looking for him, enters his village, goes to his house, kills him, and takes the tusks he has gathered. Fidow believed he had learned the necessary secrets to protect himself from the nature he exploits, but his very belief in such secrets ended up contributing to his demise. It prevented him from recognizing how identities evade full knowledge and resulted in a dangerous lack of humility. The incident confounds people in part because the elephant seems to act as would a human. Says Nonno, "How do you explain an elephant behaving as we ourselves might have behaved? . . . People cannot comprehend how an elephant could journey so far from his base, cross an international boundary, kill a man in beastly vengeance, and then leave with the remains of his massacred kith and kin" (100). The confusion of categories is further reflected in Nonno's use of "beastly" to describe the elephant's desire for vengeance after he has just claimed that the elephant acted as would a human. More generally, the elephant's actions defy the many efforts by the villagers and the press to explain them; the mystery remains.

Thus, the problem with the patriarchal pursuit of secrets is not only its genesis in the desire for mastery but also its failure to account for the fluidity, interconnectedness, and multiplicity of identity. In this novel, the effort to "know" a *defining* secret is an effort to impose one, and it leads to alienation and destruction not only of those identified as "other" but also of the self. At the end of the novel, Nonno recognizes the foolishness in trying to achieve mastery through such knowledge. Dying and blind, he proclaims to Kalaman,

> I reckon that everything has to do with the authority, the
> wherewithal to manipulate other people's destinies. In my teens,
> not humbled then, I was molded of the same clay as a madcap
> scientist . . . with too much knowledge for his own good and too
> little sense, who has the urge to remake the universe in the cast
> of his rigid formula. Power-hungry, I guessed that by replacing a
> set of magical codes with some of my own making, I might rule
> the wind and the birds which ride upon it. (298)

Nonno's summation articulates the blindness not only of his younger self but also of an anticolonial nationalist telos of another era. Farah suggests the dream of replacing colonial discourse with an indigenous set of "magical codes" that supposedly would enable one "to manipulate other people's destinies" and "rule the wind and the birds which ride upon it" was not the move from a false, arbitrary form of authority to a natural, legitimate one; instead, it was another effort "to remake the universe in the cast of [a] rigid formula" in the pursuit of power and privilege. Failing to account for the resistance of identity to a "rigid" set of defining secrets, patriarchal nationalism leads ultimately both to the destruction of the "bodies" of the nation—the land and people—and to its own undoing.

In contrast with the discourse of patriarchal nationalism, Farah represents identities as formed by a wide range of dispersed, fluid influences and by what Edward Said terms "affiliation" as opposed to "filiation." They cannot be reduced to single origins or transcendent secrets, and in fact, they evade defining representation; however, they also remain tied to the varying meanings imposed on them through their relationships. This conception of identity is reflected in Nonno's thoughts about himself:

> In a roundabout sort of way, I am asking if somebody of my age nobility, my description and disposition, has an identity outside the perimeters of the one which other persons have invented, each constructed identity having a value, the mintage of a made-up currency. To Kalaman, for instance, I am a place, a vase capable of receiving the affections with which he fills it. To Yaqut, I am the threshold of an imagined hurdle of self-appraisal. . . . In short, I am many in one, and I am *other* too. (107)

Nonno acknowledges how others shape him and his destiny through their constructions of him, and he points to the multiplicity of identity resulting from the different constructions; at the same time, he asserts, he is not defined even by the totality of these constructions and remains "other." Nonno himself may not extend the slipperiness of identity to others—it would undermine his own authority—but the novel as a whole certainly does. Farah suggests the need to relinquish notions of mastery through defining secrets and to respect the alterity that will always be there in the self *and* others. In *Secrets* such decentering of identity leads not away from ethical restraint and obligation but toward it.

Kalaman's process of maturation is defined by his movement toward this counterdiscourse. Throughout his life, he has felt uncertain about his identity and his relationships with others because of mysteries surrounding his parents' past and his birth. He has believed that if he can plumb these mysteries he will uncover a family history that will enable him to define himself clearly and know his place in the world; the secrets that have eluded him, he assumes, will give him a kind of self-mastery. However, Kalaman discovers that a key secret is his own conception through a gang rape of his mother, which means not only that he still has no idea of his biological origins but also that two of the prime, defining relationships of his life—with his father and grandfather—are meaningless, if he accepts the notion that identity is determined by one's biological origins.

As a result of his disturbing discovery, Kalaman rejects the notion that his identity and his relationships with others are defined by origins that indelibly mark his place in the world; instead, he embraces a conception of identity based on the web of influences that have shaped him throughout the history of his life. For example, he continues to view Yaqut as his father, even in the absence of a biological tie, because of the nurturing relationship they have shared: "In place of sperm, I thought it was the river of his humanity which flowed into my blood, a more precious thing, everlasting in my memory" (254). This embracing of the relationship, despite the absence of a biological connection, changes the very meaning of the word *father* for Kalaman: "The phrase 'my father' was now weighty, with moral as well as political responsibility bearing on it, notions I might not have linked to the relationship between a biological son and father" (255). Kalaman has made an important shift; he no longer thinks of family in terms of a naturally determined structure of relationships. Instead, he thinks of it in terms of a community of interdependent members formed by a history of mutual shaping. As family becomes community, as the value and meaning of others and of the self cannot be restricted to their places in a supposedly natural hierarchy, ethics and politics are no longer marginalized by assumptions regarding authentic identity and roles. They are not outside or after our private, natural selves and relationships (who we really are).

Yet Kalaman does not exactly achieve enlightenment; the notion of such closure is antithetical to the dialogic stance of the novel. Thus, for example, the degree to which he rethinks his conception of community beyond the family is unclear. Even more important, the degree to which he has questioned the kind of authority he has always associated with Nonno—benign, patriarchal, natural—remains in doubt. His notions of identity do shift, however, more than do those of Shoolongo, the target of his youthful sexual obsessions. If Kalaman felt himself to be an "outsider" in his family while he was young, she was brutally ostracized as less than human by her society. Says Nonno, "this inhumane society of ours, this our collective conscience that is the basis of our culture, has been most cruel to her as a child, as a human" (140). As a result, she comes to represent an otherness that threatens the borders of identity: "absent in her presence, present in her absence . . . Sholoongo assumes the different personae an actress assumes, while representing the full spectrum of human and animal possibilities" (200). Understandably, throughout her life she has been focused on exerting control and, to some degree, on achieving revenge, in large part by embracing her separation from social categories and norms: that is, by drawing on her reputation for having "animal power" and utilizing "her outrageous defiance of the ethos of the very society which castigated her at birth" (120).

Thus, near the end, she undoes Nonno's authority by putting him in the position of the woman in a cultural practice of a simulated rape. She achieves this reversal by manipulating what she sees as "his operating principle" of "deference to tradition" (267). In other words, like many of the patriarchal figures in the novel, Sholoongo focuses on sexual control through defining secrets. She also seeks to overturn the family hierarchy by focusing on matrilineal rather than patrilineal descent: "I thought ahead, and saw a baby, mine. . . . From Nonno I got what I was after, the correct amount of sperm having the appropriate count in strength. How relieved I was to walk . . . in the labyrinths of my capital joy, my motherhood" (274). Ultimately, however, Sholoongo relinquishes neither the underlying assumptions regarding the structure of identity nor the drive for mastery characterizing the patriarchal figures of the novel. Like them, she remains focused on instrumentalizing others and refuses any kind of ethical communal responsibility; for example, after getting what she wants from Nonno, she refuses to help him as she watches him disintegrate. Not surprisingly, she ends up in a similar position to the patriarchal characters. Her objectification of Nonno ends up creating a backlash as his sperm seems to cause in her a massive allergic reaction, and her unwillingness to rethink community leaves her without one; in the end, she can only say, "I itched. I was alone" (286).

Sholoongo was always positioned as a non- or less-than-human animal other, outside the human community. She is the most extreme example in the novel of the use of a nature/culture binary to deny the natural world and women ethical consideration and to objectify them. In this regard, Nonno reveals the gendered underpinnings of anticolonial nationalist patriarchy when he describes her as "the territory we are all tilling, a most fertile soil, dark as the good earth is yielding" (141). Yet like many early ecofeminists (for example, Judith Plant), she accepts an underlying logic of separation entailed by an essentializing identification between women and nature rather than the kind of nature/culture dynamic encouraged by most recent ecofeminists, who argue that humans' relationships with nature will always be socially mediated and who call for a sense of *solidarity* with the nonhuman.[5] Sholoongo thinks of the animal world in ideal, pastoral terms and associates it with "the maternal" and with "protective motherliness" (271). She also believes, a bit like Nonno, that in standing outside cultural codes and social relations, she enables connection with the secrets of nature. The novel questions such notions of separation and escape and represents them as extremely problematic for ecological health and the well-being of a majority of Somalia—including its nonhuman inhabitants. Throughout the novel, interactions between animals and humans—such as the episode when Kalaman and Nonno watch the Plover or when the elephant kills Fidow—suggest that exclusive conceptions of the human and the nonhuman or of culture and nature suppress their mutually mediating relationships. Farah also emphasizes that the exclusion of nonhuman others from the idea of community enables their value to be understood in strictly instrumental and monetary terms and their destruction to be treated with indifference.

Like Sholoongo, Kalaman when young identified with animals, and this identification marked a separation from the human world. However, as the episode with the Plover suggests, as he becomes older he becomes more attuned to the interpenetration and interdependence of culture and nature. He becomes aware of the ways that a significant part of Somalia's disaster has to do with environmental degradation and that deforestation, loss of biodiversity, soil erosion, and famine are interrelated: "What had been once a fertile land had now turned to fine dust, an earth as lifeless as a cut wire. Trees and forests devastated, wildlife decimated, we had a generation of farmers dead from starvation" (123). In turn, he reflects on the relationship between this disastrous situation and the arrogant belief in the ability to instrumentalize the nonhuman. Thinking of Fidow's death by the elephants, Kalaman "surveyed the scene around us and saw nothing but the signs of successive droughts. I concluded that the elephant's anger had a lot to do with man's indifference to nature, humankind's exploitative greed" (98). Kalaman seems to be moving toward a kind of awareness that decenters the human, encourages humility, and enables the imagining of, in Aldo Leopold's words, a biotic community entailing not just "privileges" but also "obligations" (203).

Yet, in a sense, the novel also brings Kalaman's biocentric formulation into question. In the case of Fidow's death, Kalaman assumes he knows the motivation of the nonhuman—"the elephant's anger had a lot to do with man's indifference to nature"—but similar interpretations of the elephant's actions by journalists are clearly problematic. They position themselves as able to explain the strange events, even though they have little knowledge of them or of the actors involved: "The world's wirelesses are broadcasting the news in as many languages as there are. . . . Some of the journalists speculate that the elephant means to give his massacred kin a decent burial. Many of the radio commentators sound triumphant. One of the local radio reporters boastfully predicts that from this day on we will have a green movement in Somalia, the first genuine one of its kind in the world" (93). The indifference to Fidow's death and the transformation of it into an abstraction points to the distancing of the journalists from those they cover, as well as their interest in making it into a sensationalistic story that will grab the world's attention. Particularly striking is the "local" effort to "boastfully" transform the incident into the beginning of the "first genuine" environmental movement run by nature. The pride these reporters take references an international competition for a purer green identity aligned with nature; in this sense, it suggests the need to approach claims to biocentric representation—which assumes the ability to identify with and speak for the interests of nonhuman others—extremely skeptically. We must always ask, the novel suggests, how social positioning and political relationships shape any representation of nature and make any claim to speak for its interests suspect.

Despite such skepticism regarding environmental representation, *Secrets* maintains a strong environmentalist ethos. It encourages the imagining of an interdependent community of humans and nonhumans shaped by social and ecological history, even as it insists on oft-unacknowledged dangers in instrumentalizing, degrading, and destroying parts of a biotic community. Farah's most recent novel, *Crossbones,* emphasizes precisely how a refusal to take into consideration such dangers by international and local actors has contributed to ever-increasing violence in and around Somalia. Taking advantage of the nation's inability to regulate coastal waters, "rampant illegal, unreported, and unregulated . . . fishing on the part of foreign fishing vessels" has put "Somalia's enormous, resource-rich maritime domain" in "danger of collapse" (Schofield 102). This situation led to escalating conflict between local fishermen and "large, technologically advanced, and often well-armed foreign fishing vessels" (109). These fishermen are labeled "pirates" by the foreign fleets, even though these fleets' poaching also constitutes a kind of piracy. Adding to the anger and fear of Somalia's coastal inhabitants, "uranium radioactive waste, lead, cadmium, mercury, industrial, hospital, chemical, leather treatment and other toxic waste" was dumped along Somalia's beaches, causing "health and environmental problems to the surrounding local fishing communities including contamination of ground water" (Eichstaedt 38). Although *Crossbones* emphasizes the responsibility of various foreign financial interests for these environmental threats, the discussions and debates among characters foreground the ways that local businessmen have also been involved. One of them started building his fortune when he both developed a partnership with an Italian fishing firm and established a frozen-food company exporting lobster. Oftentimes in the novel such characters will become involved with piracy and will strongly argue that the pirates are antiimperial warriors defending the nation's resources. Once again, Farah encourages skepticism toward a nationally tinged environmental discourse (in this case one appropriating the language of global environmental justice) even as he emphasizes socioecological violence and "the absence of food and environmental security" generated by complex transnational financial relationships and shaping a future "marked by despoliation, devastation, and more poverty" (*Crossbones* 217).

In many ways, the vision of *Secrets* corresponds to the kind of "environmental" postmodern novel touted by Dominic Head. According to Head, such a novel enables an opening (a space) for nonhuman nature—a pointing toward it and an emphasis on the importance of trying to understand and represent it—while *at the very same time* undermining the truth claims of environmental representation. In this sense, Head sees such environmental writing as aligned with postcolonialism (in its more postmodern manifestation), which tries to open up a space for the voice of the other—including within the self—while maintaining a healthy skepticism about the final recuperation of this voice. He also claims that such a novel decenters human consciousness in relation to the nonhuman natural world, even if that consciousness, in a sense, remains formally central. (It is recentered through a kind of decentering, he argues.) In *Secrets* Kalaman develops precisely through an awareness of his interdependence with the greater biosphere (the decentering of the human), through attempts to understand (and represent) nature in new ways, *and* through an (incomplete) recognition that all attempts to speak for nature will be a human appropriation of nature's voice.

Of course, a central component of *Secret*'s postmodern postcolonial orientation entails its unrelenting undermining of environmental and anticolonial pastoral narratives. In *Secrets,* narratives of the authentic and the natural enable oppressive political relationships (rather than offering escape from them) and block the development of a just, inclusive, democratic community. In this sense, Farah encourages skepticism even toward the kind of oppositional environmentalist pastoral narratives articulated by activists like Maathai.

Yet it is important to consider that Maathai's pastoral (recentering) narrative of identity could very well be a necessary strategy in some contexts for activists, writers, and activist-writers associated with struggles for environmental justice. As her use of what Rob Nixon terms "the theater of the tree" suggests, she was an astute rhetorician (*Slow* 132–37). She quickly learned how to muster international support for the Green Belt Movement and to mobilize mass resistance. In terms of her long-term goals, it is difficult to discount her contributions to the shift toward democratic change in Kenya and to increasing attention to the part played by capitalized monocrop agriculture and resource extraction in the environmental slow violence underpinning social problems in Africa. These successes were not necessarily achieved *despite* her deployment of pastoral tropes; among Kenyans, those tropes helped connect environmentalism with anticolonial struggle and refute its association with colonial conservation and with the foreign. Her stories of traditional ecological practices and values could encourage Kenyans to rally around them as signifiers of counterhegemonic collective identity and to embrace the protection of the soil as a crucial part of that identity.

In this context, we might consider potential pitfalls of an uncritical embrace of Farah's skewering of anticolonial pastoral narratives and of his emphasis on the decentering of consciousness as a locus for the reimaging of resistant subjectivity. To what degree does a focus on the individual (as productive absence) and on transformation of consciousness elide or distract from structural and ideological issues that might impinge on mobilization and the means for direct collective action? Farah's work is often discussed in terms of the effort to understand national collapse in terms of the individual "as an incarnation of Somalia" or "the patriarchal family" as metaphor (Ngaboh-Smart 130; Sugnet 740), as well as to construct a "counter-imaginary for Somalia" by reformulating individual consciousness and/or the framework of the family (Myers 139). In *Secrets* any changes that are constructed as hopeful happen primarily at the level of individual consciousness or within the framework of the family. What happens when we move this basis for hope from individual consciousness or the family unit to larger, more dispersed collectives? As Peter Hitchcock notes, in the *Blood in the Sun* trilogy (in particular) "the inevitable tendency is to read off the minutiae of character as the notional truth of nation" as if the relationship between individual consciousness and "the fate of Somalia is only one of scale" (741–42). For Hitchcock, the "particular danger" here is the risk of psychopathologizing the nation in ways that elide the complicated structural and historical causes for conditions in Somalia. In addition, *Secrets*'s positioning of a deconstructive play of identity as the ground for hope can be made to work more easily for the individual or even the family than for the formulation of counterdiscursive collectivities at larger scales, which may explain why the *Blood in the Sun* trilogy focuses on smaller scales and why so many critics emphasize the deterritorializing of the nation in Farah's work while having trouble explaining with any kind of specificity what a reterritorializing would look like.

Putting *Secrets* and Maathai's writing in dialogue in order to establish their comparative truth or universal significance would probably not be a particularly productive exercise. However, using contrapuntal reading to explore traditions of environmental writing by Africans does foster multivoiced, open approaches to environmental consciousness and action. In other words, in addition to interrogating potentially colonialist representations of African environmentalism (as belated) and pushing toward a more cross-culturally sensitive ecocriticism, it can help resist closure to environmental imagination and praxis in Africa.

3 THE NATURE OF Justice

AS A RESULT OF ENVIRONMENTALISM'S ASSOCIATION under apartheid with the priorities of a relatively affluent white minority and with racial oppression, black South Africans were often "hostile to what was perceived as an elitist concern peripheral to their struggle for survival" (Khan 15). Throughout much of the twentieth century, the state spent vast sums on wildlife and wilderness conservation and forcibly removed nonwhites from their lands in order to create national parks. Meanwhile, the majority of South Africans were left increasingly destitute, the laws of racial segregation barred them from enjoying "the country's rich natural heritage, and draconian poaching laws kept the rural poor from desperately needed resources" (McDonald, "Environmental" 257).

However, the past two decades have seen the rise of an alternative environmental justice movement that views "environmental issues as deeply . . . embedded in access to power and resources in society" and refuses to separate questions of environmental policy from problems of social injustice (Cock and Fig 16). The defining moment of change is often traced to an Earthlife Africa conference in 1992 focused on environmental justice and the subsequent creation of the Environmental Justice Networking Forum (EJNF), which grew to include over four hundred member organizations by 2000. Environmental justice in South Africa was strongly influenced by movements in the United States as a result of the parallels between environmental racism in the two countries, as well as between the civil rights and antiapartheid struggles.[1] The movement shifted the landscape of environmentalism in South Africa by linking the definition of "environmental issues" with "broader development concerns that . . . reflect relations to resources and power" (Hallowes and Butler 57). Environmental justice advocates insisted that the sites of environmental problems include the townships and homelands and foregrounded the relationship between environmental projects and the social injustices generated by the dynamics of power, privilege, and race in South Africa. As the newsletter of the South African Environmental Justice Networking Forum announced, "Environmental justice is about social transformation directed towards meeting basic human needs and enhancing our quality of life—economic quality, health care, housing, human rights, environmental protection, and democracy" (MacDonald, "What" 4).

Yet the rise of environmental justice in South Africa has by no means represented the end of green imperialism; the latter has continued under the guise of "ecological modernization." Often packaged as "sustainable development," ecological modernization focuses on the mitigation of environmental degradation through "technical and institutional solutions" and ascribes responsibility for these solutions to "scientific experts and managers" (Oelofse et al. 62–63). Its technocentric scientific orientation has been criticized as all too easily suppressing the significance of political and economic relationships and dealing inadequately "with the social questions related to assessing who benefits from and who bears the impact of development processes" (Oelofse et al. 64). Scholars and activists in South Africa have also noted the parallels between ecological modernization and colonial environmental management, which similarly sought to preserve the long-term viability of natural resources through ecologically informed management and insisted on technical rather than political framings of environmental threat.[2] Driving colonial sustainability programs was "the belief that the state and its scientists perceived natural resource use more rationally than the local inhabitants. In their eyes, this legitimized their role as ultimate stewards of the land" (Beinart and Hughes 270).[3] Those theorizing environmental justice have argued that ecological modernization in postapartheid South Africa similarly works to naturalize (neo)imperial identities and authority.[4]

Ecological modernization and, more generally, mainstream environmentalism have placed significant pressures on the environmental justice movement in South Africa over the past twenty years. "Historically white, suburban-based environmental groups" have continued to account for "the lion's share of financial resources and organizational capacity," and these groups "have been blamed for not taking environmental degradation in the townships and former homelands seriously" (McDonald, "Environmental" 260–61). Meanwhile, tensions and problems have cropped up within and among environmental justice groups in terms of the efficacy of working within hegemonic (technical and legal) channels and through the market economy. For example, research on the movement in Durban has suggested how groups can be co-opted and protest curtailed as they work collaboratively, rather than confrontationally, with industry and the state (Leonard and Pelling 143; Barnett and Scott).

According to David McDonald, among the biggest challenges faced by the environmental justice movement in South Africa is the growing relationship between environmental injustice and neoliberalism. He argues that, while environmental racism remains a significant problem, "neoliberalism and deepening class divides" pose the biggest threat to the struggle for environmental justice in South Africa:

> Policies of fiscal restraint, cost recovery, privatization, and liberalization threaten to entrench the economic disparities of the apartheid era and to drive an even deeper wedge between the "haves" and the "have-nots" of the country. The fact that many environmentally damaging neoliberal policy decisions are now being designed and imposed by a new black elite merely highlights the increasingly class-based nature of environmental politics in the country. ("Environmental" 256)

McDonald does not wish to discount the importance of race and its close relationship with class in South Africa, but he also sees neoliberalism as a particularly difficult political target given its relative abstraction and its embrace by the African National Congress (ANC). If it is true that too restrictive a focus on class can suppress the ongoing significance of structural racism, the bigger threat under the current dispensation is the protection of neoliberalism's injustices through an exclusive emphasis on race: "In the long run, neoliberalism is largely color-blind. Profit potential—not racial bias—drives its logic" (278).

In literary studies, a number of novels published since 1980 have been represented as anticipating or echoing the growth of environmental justice in South Africa.[5] Yet an incipient alternative environmentalism can be found in earlier fiction as well. Reading this fiction contrapuntally with more recent novels highlights the need to balance what was new in the environmental justice movement post-1990 with the movement's continuities with earlier struggles and with the ways those earlier struggles might be partially framed in "environmental" terms. Finally, looking at a range of novels from different historical moments, focused on a variety of places and situations, and by authors with widely divergent perspectives, serves to emphasize connections and differences among those conceptualizing the struggle for environmental justice;

it suggests that we think of "an environmental justice perspective" in South Africa not as a singular, stable concept that suddenly appeared (from elsewhere) at a particular historical juncture, but as something heterogeneous, with the flexibility to account for new challenges and to resist analytic closure.

The Place of (Neo)colonial Sustainable Development: Alan Paton and Bessie Head

Alan Paton's *Cry, the Beloved Country* (1948), Bessie Head's *When Rain Clouds Gather* (1969), and Zakes Mda's *The Heart of Redness* (2000) all develop pastoral narratives in which cosmopolitans, trying to distance themselves from a corrupt colonial modernity, establish a natural home and/or authoritative position in a rural place through their ability to catalyze the struggle against environmental degradation and, as a result, to combat poverty, communal disintegration, disempowerment, and all manner of other social ills. In this sense, these texts either reiterate or risk reiterating colonial representations of local African communities needing the leadership of European or Europeanized experts with the knowledge, skill, and sensibility necessary for understanding and protecting the land.

Yet, to varying degrees, all three novels also bring attention to the environmental injustice generated by imperialism and question a separation of conservation from a program of social transformation. In this regard, the variation among them depends on whether and how the authority of the exile is naturalized and, concomitantly, on the deployment of unitary, stable categories—nature, place, the subject—that enable the process of naturalization.

The potential alignment between an environmental justice framework and *Cry, the Beloved Country* is stunted by Paton's inability to relinquish colonial categories. As Lewis Nkosi so astutely observed, *Cry, the Beloved Country* is focused on the fears and desires of the white liberal in the years preceding apartheid. Its central, guiding sensibility is not Stephen Kumalo, the naïve black minister who travels to the city for his brutal "education" and then returns a wiser man, and not even John Jarvis, the white farmer who also must learn of the realities of South Africa through his travels to Johannesburg, but the narrator, who already has learned those realities and embraced a liberal vision for South

Africa's future. He recognizes the disastrous social consequences and moral bankruptcy of the country's racial policy but fears the implications of black nationalism and revolutionary rhetoric that could lead to a new dispensation. He desires a shift from the particular manifestation of colonialism that is developing in South Africa but cannot relinquish a colonial conceptual ordering of the world and the attendant authority it accords him. In line with its realism, the novel mostly endorses the narrator's perspective and, ultimately, offers a resolution to the white liberal's dilemma through individualist and technical solutions to South Africa's problems, which are closely tied to colonial conservation. Yet the contradictions embedded in this liberal solution threaten to undermine it and, ultimately, cannot be contained. Near its conclusion, the novel points to its inadequacies and gestures toward the need for a different approach if the environmental degradation that the novel represents as the driver for social ills is to be addressed.

Cry, the Beloved Country embraces a romantic conservationist ethic. From the first chapter, the land in its natural state, "being even as it came from the Creator," is represented as embodying a transcendent, "holy" spirit reflected in a beauty, "beyond any singing of it," that outstrips human language and in a bounty that "keeps men, guards men, and cares for men." This beauty and bounty are enabled by a perfect natural system for irrigation and soil retention: "The grass is rich and matted, you cannot see the soil. It holds the rain and the mist, and they seep into the ground, feeding the streams in every kloof." If the land is not cherished and protected, if its ecological relationships are not conserved, then God's grace is withdrawn: "destroy it and man is destroyed" (33).

This warning sets up the novel's representation of environmental degradation as a central cause for South Africa's social problems. As one moves from "the rich green hills" to "the valley below," a classic scene of soil erosion unfolds in which the "red and bare" land "cannot hold the rain mist, and the streams are dry in the kloofs" (33). This devastation results from a lack of care: "Too many cattle feed upon the grass, and too many fires have burned it" (33–34). As more soil is lost with each rain, as the earth is "torn away like flesh" and the streams run full of "red blood," God's spirit is murdered. The people lose its protection and can no longer sustain themselves, the "young men and girls" leave for the city (34), the community falls apart, traditional values are lost, those who are left behind become increasingly destitute, and those who leave experience the deprivation and degradation of the city.

The relationship between this (clichéd) pastoral narrative and colonial ideology is suggested by the contrast between the representation of poor "native" environmental practice and the depiction of agriculture practiced in the hills, where the grass is "well tended" (33). Such proper stewardship is associated with the narrator and those like him, who have the requisite environmental reverence and knowledge to "keep" and "guard" the earth. This dynamic becomes explicit when, at the beginning of the second "book," we are introduced to the character of James Jarvis, who lives among the hills in "a small and lovely valley" and has "one of the finest farms of this countryside" (161). Jarvis has the kind of relationship with the land that allows for its beauty and health to be maintained. The condition of the farm and its surroundings—as well as its location, "high above Ndotsheni, and the great valley of the Umzimkulu" (161)—marks the settlers' superior ecological knowledge, appreciation of God's spirit as embodied in nature, and breadth of vision, which those who live in the "reserves" of Ndotsheni lack: "There were too many cattle there, and the fields were eroded and barren. . . . Something might have been done, if these people had only learned how to fight erosion, . . . if they had ploughed along the contours of the hills. But the hills were steep, and indeed some of them were never meant for ploughing. . . . And the people were ignorant, and knew nothing about farming methods" (162). Historically, such representations of Africans causing soil erosion through unsound "farming methods" such as vertical (rather than horizontal) plowing and overstocking became a central component of colonial ideology in African settler colonies.[6]

These representations underpin one of the novel's central solutions to South Africa's problems: sound agricultural practices and effective conservation measures in the reserves developed by whites who have adopted a selfless missionary spirit. Jarvis becomes the emblem for this solution as he is transformed through the tutelage of his martyred, liberal son's writing. Arthur Jarvis argued that the admirable goals of colonial development have in South Africa been perverted by "selfishness" and that part of this development must entail the difficult work of setting up a "system of order and tradition and convention" that could replace the moribund "tribal system that impeded the growth of the country" (179). Taking his son's philosophy to heart, Jarvis begins a number of selfless conservation and development projects in Ndotsheni when he returns to his farm. Depleting his own economic resources, he funds and supervises the building of a dam and hires an agricultural demonstrator to instruct the people on proper environmental practice, including horizontal plowing, reduction of cattle stock, differentiation between proper and improper agricultural land, and tree planting. For these projects to be successful, ignorance, narrow self-interest, fear, and misguided tradition among the "natives" must be overcome. The process is difficult, but it eventually yields positive results as the land begins to return to health. In turn, the famine and general lack of resources that drive people to the corrupting city are alleviated. Thus, through a gradual, local, and technical solution based on individual change of heart and action, the problems of environmental degradation and rural poverty raised by the novel are addressed without sudden, drastic change and without political action. At the same time, the authoritative position of the settler, based on his understanding and close connection with the spirit of the land, is reinforced.

In many ways, the representation of Jarvis's projects reinforces the ideology of colonial conservation in Paton's time. In the 1940s, such conservation was focused on "instruction" in proper agricultural methods and development projects such as dams. It did not yet have the same punitive elements that would develop under apartheid; however, it was extremely hierarchical and interventionist, seeking no input from local people (Beinart, *Rise*). It assumed that Africans were unenlightened and lacked the objectivity embedded in science and enabled by the European temperament necessary for sustainable development. As a result, resistance to conservation was represented as resulting from Africans' inability to transcend tradition, childlike narrowness of vision, and self-interest. Environmental degradation and any failures in colonial conservation are always the result of Africans themselves rather than of colonial policy produced by disinterested, enlightened colonial scientists, policy makers, and administrators.

The text's depiction of Jarvis's dam project is an especially powerful example of the perspective entailed by colonial conservation. Kumalo (a clear leader in the community) first learns of it when he observes Jarvis, the "Magistrate," white surveyors, and the "chief" meeting to lay the ground work for the dam. Kumalo is depicted as unable to understand what he observes; he "stood more and more mystified" (276). Meanwhile, the chief's efforts to help are represented as childlike in their ignorance; he becomes a nuisance as he tries to display his authority by helping to distribute the "sticks" in the ground. The scene legitimates a relationship in which Europeans consult among themselves regarding what is to be done while Africans are relegated to the positions of passive recipients of white knowledge, skills, and largess. Far from questioning the lack of input from the community, it suggests that any such input would be ignorant and foolish. The potential problems with authoritarian conservation are further contained later by the success of the projects and by the suggestion that any unhappiness with the new policies comes from "custom," fear, and narrow self-interest.

Much of the history of South African conservation in the first half of the twentieth century is suppressed by such representation. Far from being objective, conservation science and policy were shaped by colonial economic and political priorities and by the prejudices embedded in colonial ideology. As a result, they were an integral part of policies and practices that were not only socially but also environmentally destructive. Representations of ignorant, destructive indigenous agricultural practice were used to justify land alienation, which greatly exacerbated environmental degradation as more and more people were forced to eke out livings from less and less land. Furthermore, resistance to the transformations of "systems of agriculture, under the banner of 'betterment,'" resulted both from their link with efforts to extend colonial control in the reserves and from the "destruction of community and collective practices and identities" that they entailed (Beinart and Hughes 284, 288).

Ultimately, the text cannot contain the problems implicit in either liberalism or colonial conservation. In a discussion near the end between Kumalo and the young agricultural demonstrator, the latter suggests that the gradualist liberal solution based on education, technical projects, and individual changes of heart is not adequate; colonial land distribution must be reversed, and it must be done soon. The demonstrator argues that land alienation (which by the 1940s had put 87 percent of the country in white hands) has left too little land to support the populations in the reserves and severely attenuated the significance of any local improvements. He notes that "it was the white man who gave us so little land, it was the white man who took us away from the land to go to work," and then rhetorically asks, "if this valley were restored, as you are always asking in your prayers, do you think it would hold all the people of this tribe if they all returned?" Finally, he proclaims, "we can restore this valley for those who are here, but when the children grow up, there will again be too many" (302). Kumalo cannot respond to the demonstrator's argument—and neither can the text: "To this young man Paton allots the last and most telling word. To his logic Kumalo and his patron Jarvis, with their fragile hope of preserving an Eden in the valley immune from the attractions of the great city, have no response" (Coetzee 129). The demonstrator suggests the need for a wider, political solution enabling land redistribution, as well as the impossibility of separating questions of environmental degradation from economic injustice and colonial history. In this sense, because the text does not really have any response to the demonstrator, his voice would seem to open up a space for a discourse of environmental justice as a solution to the problems Paton depicts.

This space certainly remains circumscribed by the text's liberalism. In fact, even in this conversation, fear of a political solution is evoked and managed when the demonstrator takes the position—echoed by all of Paton's other "admirable" black characters—that in order for positive change to occur politics must be eschewed. Concomitantly, the text shuts down a more democratic, egalitarian approach to conservation and development; whites must teach and lead, blacks must listen and follow. In fact, the demonstrator reiterates colonial assumptions regarding knowledge, development, and conservation when he claims that even his "love for truth" has been "taught" to him "by a white man," while also noting that "there is not even good farming . . . without the truth" (302–3).

However, if Paton cannot entertain a revolutionary political solution and if he cannot relinquish the identities and relationships bequeathed by colonialism, he also has no answers to the limitations of liberalism and colonial conservation, which he articulates through the demonstrator's arguments regarding land distribution. Stuck between an unwillingness or inability to move beyond those solutions and their limitations, the text can only end uncertainly, with a need for a Christian faith in the future.

In some important ways, the representations of conservation and development in Bessie Head's *When Rain Clouds Gather* are substantially less colonialist than those in *Cry, the Beloved Country*. Success depends on a relationship between outsiders and Indigenes that is not as hierarchical and is more participatory than the kind of relationship assumed necessary by Paton. In order to bring "progress," those outsiders must live with and like the people, and they must listen to and take seriously local voices. Concomitantly, Head's novel does not endorse development and conservation projects conceived and imposed from outside the local community. The goal is not the transcendence of people's relationship with place but a movement forward within the parameters of that relationship. However, ultimately, *When Rain Clouds Gather* is neo- rather than postcolonial. Head may not try to resuscitate stereotypical colonial relationships, but she does reproduce models of geographic and collective identity, of authority and of knowledge—and, as a result, of environmental stewardship—entailed by colonialism.

When Rain Clouds Gather focuses on two intertwined challenges: How can independent, strong-willed exiles from restrictive, artificial societies find belonging and their rightful places in a new, more natural home? How can a place be developed in ways that will evolve from and preserve its unique environmental and cultural features? The solution to these challenges, embodied in the utopian village of Golema Mmidi, is a form of bioregionalism. At the heart of bioregionalism is the concept of reinhabitation, which involves "a commitment to understand local ecology and human relationships" (Thomashow, "Toward" 125), and a notion of belonging based on the belief "that as members of distinct communities, human beings cannot avoid . . . being affected by their specific location, place, and bioregion" (McGinnis 2). The bioregion itself is an ecopolitical unit integrating "ecological and cultural affiliations" and "derived from landscape, ecosystem, watershed, indigenous culture, local community knowledge, environmental history, climate and geography" (Thomashow, "Toward" 121). If capitalism in its various phases has made space out of place, stripping away prior signification (deterritorializing) and reshaping in order to facilitate control and exploitation, then the process of imagining or reimagining a "place" entailed by bioregionalism can be one means of countering threats of exploitation, environmental degradation, and disempowerment. This form of imaginative engagement encourages understanding of and commitment to one's place and community, as well as the places and communities of others, and it challenges predominate geopolitical forms of identity and the atomization encouraged by capitalist modernity.[7]

In *When Rain Clouds Gather,* Head depicts the area around Golema Mmidi as a kind of bioregion with distinctive, interconnected cultural and ecological markers. It is a natural geographic unit, defined by a cohesive spirit in the land that has shaped the people's consciousness and their culture; in this sense, it brings into question the abstract political and geographic units made by humans—nation, tribe, race, and so on.

In the novel, movement forward must entail an appreciation for and desire to preserve the spirit of the bioregion. In this sense, the novel is in tune with the underlying moral principles embraced by many bioregionalists—preservation and sustainability.

This bioregionalist sensibility is especially apparent in the novel's treatment of exile. Huma Ibrahim reads *When Rain Clouds Gather* as propelled by what she calls "the exilic consciousness," which seeks for a sense of belonging in a new land and a transformed self free from the debilitating effects of her former oppressive home. This reading coincides with a common "biographical" interpretation in which the novel is read as Head's own symbolic means of achieving belonging in Botswana and of escaping South Africa—that is, moving beyond its colonial, racist boundaries: "the strongest desire is the exile's longing to make herself new—to be born again in a different nation" (63). In order to depict the needed conceptual separation, Head portrays the natural spirit of the area around Golema Mmidi as the antithesis of the artificial, debilitating society from which the exiles come and offers the potential for healing, self-realization, and a sense of true belonging. However, for this potential to be realized, the exiles must establish their places in the new home. On the one hand, this involves understanding, identifying with, and working within the parameters of the new environment, as well as shedding the psychic and ideological limitations they bring with them. On the other hand, belonging necessitates establishing their particular roles in the community. In *When Rain Clouds Gather,* these roles stem from the exiles' ability to foster progress by getting in touch with the spirit of the land in a way that the Indigenes cannot. The experience they bring, their temperaments, and their exilic desire and distance enable them to avoid and combat the restrictive aspects of local culture and become catalysts for a form of development that stems from reinhabitation.

The political implications of the novel's bioregionalist sensibility are complicated. In general, bioregionalism tends to challenge some dominant models of collective identity (in particular nationalism) and to encourage suspicion of externally imposed development. With its focus on reinhabitation, it can initiate the search for place-based models of development based on the valuing of local communities, cultures, and natural environments. At the same time, bioregionalist belonging can all too easily reinforce reactionary, xenophobic forms of consciousness. Notions of bounded places threatened by external forces can support hostility to outsiders and a restrictive notion of who belongs. They can also erase the effects of history and, as a result, naturalize specific constructions of place and identity and their political inflections.[8]

When Rain Clouds Gather certainly reflects the antiimperial tendencies of bioregionalism. It suggests that any effective progress benefiting local places and people comes from the knowledge, appreciation, and close relationship entailed by reinhabitation and dialogue, rather than from grand development schemes imposed by external authority. In addition, it avoids the xenophobic tendencies of some forms of bioregionalism through a focus on the positive influence of exiles and of boundary crossing. This focus has been praised by critics, who argue that Head offers a kind of counterhegemonic narrative that breaks from the discourses of patriarchy, racism, colonialism, and tradition.[9] Yet the novel still embraces a certain idealist, bounded conception of bioregion and, more generally, place and identity; this idealism, which suppresses the mediation of knowledge and representation by social position and culture, ends up making natural many of the assumptions regarding identity, power, and knowledge underpinning colonial conservation.

In many ways, *When Rain Clouds Gather* is a fairly straightforward romantic pastoral narrative. The two central exiles, Makhaya and Gilbert, find in Botswana a natural, peaceful, ordered home where they can escape the damaging effects of an irredeemably repressive South Africa and an oppressive, bourgeois Britain. As a result of the "great goodness" to be found in Botswana, Makhaya is able to dampen the impact of apartheid and of being born into "one of the most custom-bound and conservative of tribes in the whole African continent" (181, 120). Both colonialism and ossified traditionalism have dehumanized and alienated him; as a result, he has never felt a sense of connection to community or place.

The morning after he crosses the border from South Africa and steps onto "free ground," Makhaya sees his first dawn in Botswana and already feels on the path to an authentic sense of belonging: "he simply and silently decided that all this dryness and bleakness amounted to home and that somehow he had come to the end of a journey" (4, 11). The novel has many such moments, in which the particular beauty of Botswanan nature gives Makhaya a feeling of connection and helps to heal him. In a later scene, the narrator reflects on how Makhaya's growing relationship with the landscape enables him to achieve a sense of wholeness and to overcome the trauma of his past. During July in

Botswana, "filtered light covered everything with a glossy, soft sheen," and "mysterious blue mists clung like low, still clouds, all day long, on the horizon": "Makhaya found his own kind of transformation in this enchanting world. It wasn't a new freedom that he silently worked towards but a putting together of the scattered fragments of his life into a coherent and disciplined whole" (118).[10] The contrast between Botswana and South Africa is also marked by the challenge the Botswanan characters face in trying to understand Makhaya; their environment "was one full of innocence. The terrors of rape, murder, and bloodshed in a city slum . . . were quite unknown to Dinorego, but he felt in Makhaya's attitude and utterance a horror of life, and it was as though he was trying to flee this horror and replace it with innocence, trust, and respect" (93). Particularly absent from the Botswana of the novel is the perversion of nature represented by South Africa's particularly virulent, dehumanizing form of colonial racism, in which "black men . . . were like Frankenstein monsters, only animated by the white man for his own needs. Otherwise they had no life apart from being servants and slaves" (129).

Despite this romantic representation, Botswana is not represented as Edenic; the narrative is propelled by the efforts of the exiles in Golema Mmidi, especially Makaya and Gilbert, to break down the various forms of tradition that truncate peoples' identities and prevent them from leading more prosperous lives. The people and the land need the exiles to catalyze positive change and help them achieve their potential and become more fully themselves. Because of their willingness to challenge tradition and to cross various kinds of boundaries, as well as the fact that they bring new perspectives and new forms of knowledge, the exiles are crucial for the development of liberating, more productive forms of community and dwelling. Thus, the many "outsiders" who live in the village of Golema Mmidi make it a center of progress. As the narrator says at the end of the novel, this is a village where "the Good God" had brought together "all his favorite people" in order for "them to show everyone else just how quickly things could really change" (184).

Golema Mmidi represents a way of dwelling that breaks from local tradition and enables closer environmental connection. The very name of the place marks its departure from a communal identity defined in terms of customary social and political connection to one identified by a prioritizing of the relationship with the land.[11] Golema Mmidi's name "marked it out from the other villages, which were named after important chiefs or important events. Golema Mmidi acquired its name from the occupation the villagers followed, which was crop growing. It was one of the very few areas in the country where people were permanently settled on the land" (16). Golema Mmidi can be described as the product of a kind of bioregional reinhabitation that is, ironically, spurred by the condition of exile. The effort to establish a home for the self necessitates a conscious search for environmental connection and leads to permanent settlement on the land; "necessity . . . rejection and dispossession" had forced the exiles of Golema Mmidi "to make the land the central part of their existence. Unlike the migratory villagers . . . they built the large, wide neatly thatched huts of permanent residence" (16). In the novel, the "migratory" culture associated with the natives is represented as environmentally destructive and, consequently, leads to gradual decline in quality of life. Livestock strip the land of vegetation, causing soil erosion, and there is little understanding of the damage being done or ways to reverse it. In contrast, the exiles' environmental knowledge and appreciation enable effective stewardship and agricultural development.

The novel's suggestion that exiles are crucial for the development of better, alternative ways of dwelling on the land comes into particular relief through Gilbert's character. He comes to Botswana in order "to assist in agricultural development and improved techniques of food production" using his scientific expertise (17). Toward the goal of changing "unsound agricultural practices" and "primitive techniques that ruined the land," he sets out to weaken rigid "traditions" and develop an understanding of local ecology (29, 17). Because of the destruction of local biodiversity from overgrazing, he tries various experiments to learn how indigenous plants germinate, spread, and interact with one another. Through his efforts, Gilbert gains some insight into the original ecosystem: "Had all this strange new growth lain dormant for years and years in the soil? He questioned the villagers, but only Dinorego, one of the earliest settlers, retained a wistful memory of when the whole area had been clothed by waist-high grass and clear little streams had flowed all the year round" (32). In turn, Gilbert's newfound knowledge enables him to develop environmental practices that hold out the promise of a return to this richer, healthier originary ecosystem.

For Gilbert and other exiles to bring "progress," however, they must also develop an understanding of and connection with the people. From living like and with the locals, Gilbert begins to see the world differently, until "his whole outlook was entirely Botswana" (110). This "outlook" enables him to see how the arid, desertlike environment has shaped the people's consciousness, including their conceptions of themselves, their relationship with the land, and their ability to move forward. Gilbert's understanding of their flight before the "expansive ocean of desert" and subsequent contentment "to scrape off bits of living from its out-skirts . . . gave him a different outlook on subsistence farming. If a man thought small, through fear of overwhelming odds, no amount of mod-ern machinery would help him to think big. You had to work on those small plots and make them pay" (111). Gilbert's growing recognition that change must "follow the natural course of people's lives rather than im-pose itself . . . from on top" is coupled with an appreciation for what is worth preserving, not only in the environment but also culturally (25). For example, through Dinorego he comes to learn of the ways that "a Batswana man" is flexible, "ready to try a new idea," and to appreciate how such a man learns through observation and example (19–20). When the text indicates a problem with Gilbert's perspective, it is precisely when he does not think in terms of local conditions and, instead, tries to place Golema Mmidi's progress in an abstract grand narrative originat-ing elsewhere. At points, Gilbert vaguely gestures toward a classic cap-italist developmental vision in which agricultural improvements and, especially, the growing of tobacco as a cash crop will lead to a West-ern "advanced" lifestyle for the inhabitants of the village. However, the novel undermines this vision, in part by highlighting the value of what would be lost—including the morally and spiritually charged natural landscape. Near the end of the novel, the narrator proclaims:

> What was [Gilbert] looking for? What was he doing? Agriculture?
> The need for a poor country to catch up with the Joneses in the
> rich countries? Should super-highways and skyscrapers replace
> the dusty footpaths and thorn scrub? It might be what he said he
> had in mind. . . . But the real life he had lived for three years had
> been dominated by the expression of Dinorego's face, and God
> and agriculture were all mixed up together after these three years.
> (181)

This God to which the narrator refers lives "in the small brown birds of the bush" and "in the dusty footpaths" (182). When Gilbert situates Golema Mmidi in capitalist modernity's developmental narrative, he temporarily loses touch with the transcendent, redemptive spirit em-bedded in the inhabitants and environmental features of this place—a spirit that, the novel suggests, any legitimate model of progress should conserve.

This romanticism is a crucial feature of *When Rain Clouds Gather*'s "green" sensibility. Yet Head's concerns with conservation are always also related to issues involving people's livelihoods and empowerment (or disempowerment). The value of nature, including biodiversity, is closely tied to its significance for the community's economic well-being and ability to resist oppressive political relationships. The narrator even suggests that Gilbert's occasional drift toward a focus on nature for its own sake, apart from human concerns, is a flaw fueled by a desire to avoid the complications of the human world. The novel's concern with the relationship between conservation and justice is highlighted by the role of women in Gilbert's efforts to reform agricultural practice. Women have been primary leaders and participants in movements associated with the environmentalism of the poor, a centrality often explained in terms of their positions in rural communities as those most impacted by and attuned to environmental degradation.[12] Recognizing that the village's women are most likely to act against environmental degradation, Gilbert mobilizes them to make changes and empowers them, both by giving them more control over their resources and by investing them with knowledge and authority.

At the same time, Head's depiction of Gilbert reinforces many of the assumptions underpinning hegemonic colonial conservation. The novel suggests that the combination of his scientific expertise, his careful attention to local conditions, and his willingness to challenge existing environmental practice enables him to grasp ecological conditions accurately and to institute sound environmental practices. In contrast, time, local people and culture are represented as ecologically ignorant and destructive. Their consciousness is determined by custom and by the local environment to such an extent that they cannot, on their own, foster the proper objectivity and perspective that will enable sustainable development. For that, they need Gilbert. At one point, the novel even uncritically parrots the notion that Africans lack an adequate environmental sensibility and that, as a result, they need the guidance of Western culture and science. In a moment of apparent insight, Makhaya thinks, "Gilbert's culture [was] one that had catalogued every single detail on earth with curiosity, and it revealed to him . . . how impossible it would be for Africans to stand alone. His own culture lacked, almost entirely, this love and care for the earth" (131).

Such celebration of Western environmental knowledge and sensibility, combined with Head's denigration of local livelihood practice, gives a troubling inflection to the use of *When Rain Clouds Gather* as an introduction to Botswana for "Peace Corps and other international aid workers" (Nixon, *Homelands* 111). Coreen Brown argues that the novel was "suggested reading" for these workers because it provided "an accurate account of farming procedures" (54).[13] Such a claim is highly problematic; as Jonathan Highfield trenchantly argues, Head ignores "the negative effects caused by the importation of agricultural techniques and foodways" and embraces a "vision of agriculture" that is "unsustainable in as dry an environment as southeastern Botswana" (116–17). Rather than necessarily suggesting the accuracy of Head's representations, aid organizations' use of *When Rain Clouds Gather* might plausibly be connected to the ways the novel reinforces discourses about expert knowledge and aid in Africa that have legitimated imperially tinged forms of external intervention.

The forms of identity normally associated with colonialist representation would seem to be attenuated by Gilbert's need of the help, even guidance, of African characters—and particularly Makayha. In the novel, sustainable development cannot happen without their input and leadership. However, these characters are all exiles catalyzing change within a local culture that lacks an internal historical dynamic. Ultimately, Head may to a degree challenge colonialism's racism, but she still generates what might be termed a neocolonial bioregionalism in which African elites work with foreign experts to establish a form of development guided by and protecting the identity of a place. The local people are closely identified with the land and are passive carriers of its natural spirit, while the exiles catalyze change that enables a further unfolding of that spirit. In this sense, while *When Rain Clouds Gather* may be (in alignment with bioregionalism more generally) skeptical about nationalism itself and may challenge the gendered roles implicit in much postindependence nationalist discourse, Head's form of bioregionalism does not do away with those roles but assigns different identities to them. Instead of women being the carriers of tradition, local people and environment are the vessels for a natural essence of place, and elite outsiders of both genders occupy the male role of modernizers who are also stewards of that essence.[14]

A crucial component of Head's neocolonial bioregionalism is the cleansing of a place's essence from external influence or the shaping effects of history. Her idealism entails a conception of the bioregion as having an autonomous, stable identity that can ground the right to power. As the exile connects with the spirit of the land in ways that enable better environmental knowledge, appreciation, use, and care than can be supplied by those already living on it, the exile's authority is sanctioned by nature. In contrast, if an originary, transhistorical geographic identity was put into doubt, then representations of authority in place would be more unstable and contested. At the same time, the ability of the exile to access and accurately represent the "true" natural essence of place requires transcendence of the influences that have shaped him or her; if this transcendence is brought into question, the purity of the exile's knowledge of and connection with the land is as well. At the nexus of a coherent, apolitical bioregional identity and transcendent knowledge, the exile is able to establish a natural home and authority for the self, free from the distortions of both imperial ideology and slavish tradition.

The Place of Postapartheid Sustainable Development: Zakes Mda

If the potential for a perspective aligned with environmental justice is stunted in *Cry, the Beloved Country* by Paton's inability to relinquish colonial identities and if in *When Rain Clouds Gather* it is circumscribed by Head's incipient neocolonialism, it would seem to be more fully realized in *The Heart of Redness*. Mda mostly undermines colonial conservation's heroic narrative; for example, he skewers the contrast between, on the one hand, knowledgeable outsiders with the skills and vision to protect the land and move the community forward and, on the other hand, naïve, simple, and ignorant locals. More generally, *The Heart of Redness* evinces skepticism toward the dichotomies and unitary identities embraced by Paton and Head. Yet it cannot offer a path for action that breaks from the reformist principles of ecological modernization. Its seemingly successful solution to the environmental and economic threats to a community's well-being remains overly reliant on local strategies and on the form of capitalism that causes those threats in the first place. In other words, that solution encourages a perspective that separates an increase in environmental justice at the local level from political and economic processes operating at other scales. The result is a potential focus on the protection and empowerment of one's own place and the deprioritizing of resistance to—even the un-intended reinforcement of—those processes that continue to destroy other places and that always threaten any temporary victories at the local level. Yet, ultimately, the novel draws attention to the limitations of the supposed solution and, in the process, resists the kind of conceptual closure that can threaten the vibrancy and effectiveness of environmental justice struggles in the long run.

The Heart of Redness switches between a story of postapartheid South Africa and an interconnected narrative depicting events from the period of the Xhosa cattle killing in the 1850s. Mda's recounting of these events relies heavily on J. B. Peires's *The Dead Will Arise* (cited in the novel's dedication as its primary historical resource).[15] Since the eighteenth century, the amaXhosa had increasingly been dispossessed as the British expanded the size of the Cape Colony north and east. The colonial land grab was accompanied by the destruction of the amaXhosa's political and economic independence and the reshaping of their religious and cultural institutions. Efforts at resistance had not only failed but enabled quicker colonial access to the best land. In addition, during the mid-1850s over half of the amaXhosa's cattle were wiped out by lung sickness caused by a pathogen introduced by imported cattle, and they endured a crop blight that ruined their maize. In the midst of this tragedy, the prophetess Nongqawuse claimed to hear the voices of the ancestors telling her that the people must kill all their cattle, destroy their grain, and refrain from agriculture to purge impurity from the land and enable the rising of the dead. The ancestors would bring with them clean animals and grain, and the English invaders would be swept away. The people became divided between believers of the prophecies and unbelievers, resulting in civil war. The results of the cattle killing were apocalyptic for the amaXhosa. Over eighty thousand starved to death or were forced by starvation into the colonial labor market. The colonial governor, Sir George Gray, turned the tragedy to British advantage, taking hundreds of thousands of acres of Xhosa land for the use of white settlers. Even Peires struggles to articulate the catastrophic aftermath: "The impact on the Xhosa . . . is difficult to express in words. Their national, cultural and economic integrity, long penetrated and undermined by colonial pressure, finally collapsed" (321). Yet, while in no way mitigating the apocalypse, *The Heart of Redness* emphasizes the survival, if also transformation, of many Xhosa institutions.

In the novel's postapartheid present, a local, marginalized community faces another kind of imperial threat: environmental apocalypse by a tourist development scheme driven by the combination of predatory capital and a distant national government. The scheme involves a casino resort to be built near the Xhosa seaside village of Qolorha in the Eastern Cape by "a big company that owns hotels throughout Southern Africa" (66). This plan suggests the continuation of apartheid-era tourist development strategies. The hotel giant Sun International began construction of the Wild Coast Sun Casino in 1979, and before it was completed in 1981, "103 households were forced to move" without compensation (Dellier and Guyot 83). It was, and still is, a classic example of the fenced resort in a beautiful but poor area; local communities, which have seen no benefit from the casino, are kept out while wealthy guests are "localized inside the enclave" (Simukonda and Kraai 53; Dellier and Guyot 82).

In *The Heart of Redness,* many of the villagers initially support the casino scheme, believing it will bring "jobs, streetlights, and other forms of modernization to this village" (67). The novel suggests that this dream of modernization is a relatively worthless commodity being sold to enrich national elites at the expense of the local people and environment. The casino will bring few jobs for the local people, even as it takes away their access to land and resources. The novel focuses especially on the ways that this particular form of tourist-centered development will obliterate the environment: "a project of this magnitude cannot be built without cutting down the forest of indigenous trees, without disturbing the bird life, and without polluting the rivers, the sea, and its great lagoon" (119). The ultimate threat represented by the casino scheme is revealed when the developers come to talk to the people of the village. By the end of the "discussion," the two men are ignoring their audience and debating the virtues of building the casino and hotel—complete with an amusement park—versus those of making "a retirement village for millionaires" with "trees imported from England": "We'll uproot a lot of these native shrubs and wild bushes and plant a beautiful English garden" (202–3).

However, the postapartheid story foregrounds another kind of outsider, Camagu, a Xhosa man who has lived most of his life in the United States, where he received his PhD in communications and development studies, and who returned to South Africa to help build a postapartheid nation. Unable to find a job, because he is not one of the "aristocrats of the revolution," he becomes disillusioned with the new dispensation. Chasing after NomaRussia, a girl he meets at a funeral in Johannesburg, he goes to Qolorha and eventually espouses the fight against the casino. He comes up with a plan to thwart it through the designation of Qolorha, the place of Nongqawuse's original prophecies, as a national heritage site and through the development of a backpacker hostel. Ultimately, as was true with the exiles of *When Rain Clouds Gather,* Camagu seems to establish a position for himself in his new rural home, Qolorha, in part through his ability to foster sustainable development.

As was also true of Head's novel, *The Heart of Redness*'s conceptual grounding for a solution both to environmental threat and to the exilic condition is bioregionalist. Yet Mda avoids, more so than Head, the potential pitfalls of a bioregionalist outlook. Central among these pitfalls is what Rob Nixon refers to as ecoparochialism, the threat that the focus inward toward a bioregion will result in occlusion of the ways that any region is established by relationships with other places and has permeable boundaries: "All too frequently, we are left with an environmental vision that remains inside a spiritualized and naturalized national frame" (*Slow* 238). As Nixon implies, bioregionalists can be overly confident in their ability to achieve representational closure. One example involves the very definition of the bioregion itself, which, as Lawrence Buell has pointed out, often elides the "malleable or problematic" nature of a bioregion's borders (*Future* 84). The potential lack of productive skepticism in bioregionalism can extend to its conception of environmental protection. Doug Aberley claims that the decentralization entailed by the bioregion enables "the achievement of cultural and ecological sustainability" (37). However, the principle of sustainability—"of more prudent, self-sufficient use of natural resources"—is no guarantee, since it "requires guesswork about what future generations will be like" and "runs contrary to the known fact that nature itself does not remain stable" (Buell, *Future* 84). Finally, the notion of the "bioregion," like any bounded notion of place, can all too easily lend itself to xenophobia with its attendant nativism. For those concerned with processes of displacement resulting from the history of imperialism, this potential xenophobia can be especially disturbing (Buell, *Future*).

The approach to place offered by *The Heart of Redness* is most closely approximated by Thomashow's concept of *cosmopolitan* bioregionalism, which significantly departs from more traditional versions of bioregionalism. According to Thomashow, cosmopolitan bioregionalism would recognize how a "local landscape" cannot be understood without reference to larger systems, as well as how the belief in hard boundaries "is the cause of much human suffering." It would embrace the notion that borders are "permeable" and identities "pluralistic" and "multiple" ("Toward" 129). For the cosmopolitan bioregionalist, the bioregion is produced as much by the instability and dislocation of "ecological and cultural diasporas" as by indigenous nature and culture: "A bioregion is the stopping place for the migration of assorted flora and fauna, each of which makes its indelible imprint on the ecology and culture of the neighborhoods where they temporarily reside" ("Toward" 123, 129). Ultimately, the kind of "bioregional sensibility" Thomashow seeks "requires multiple voices and interpretations" and is "necessarily open-ended and flexible" ("Toward" 130). Like Thomashow, Mda emphasizes the need for an approach that can account for shifting and multiple constructions of place and identity. In *The Heart of Redness*, representations of place are always unstable both because of change over time and because of a lack of unity at any one time; they need to be recognized as human constructions open to debate in order to be effective in protecting the local. Ultimately, Mda's novel can help develop a cosmopolitan bioregionalism not only by emphasizing the instability inherent in efforts to define a local identity but also by suggesting how such efforts will always be politically inflected.

Nixon calls for just this sort of revision of bioregionalism as part of the larger project of bringing together postcolonial writing and ecocriticism. An ecocritical "ethics of place," especially as embodied in bioregionalism, could be very helpful for postcolonial critics theorizing the threats to local environments posed by imperial processes and the grounding for resistance. However, mainstream environmentalism is all too often inflected with a focus on purity ("virgin wilderness and the preservation of 'uncorrupted' last great places"), with national (as opposed to cosmopolitan) frameworks, and with the suppression or subordination of history. As a result, it has encouraged "spatial amnesia" in which the histories of indigenous peoples and of the making of places in transnational and colonial contexts are suppressed (*Slow* 236, 238). A central part of the postcolonial ecocritic's task, Nixon argues, is to take up the "intellectual challenge" of drawing "on the strengths of bioregionalism without succumbing to . . . ecoparochialism" (*Slow* 239). *The Heart of Redness* affords an ideal opportunity to explore how these "strengths" might be mobilized in efforts to resist imperial development *and* how the concept can be transformed in the process.

In its pursuit of a cosmopolitan bioregionalist vision, *The Heart of Redness* upholds struggles to defend the indigenous and the local; yet, at the same time, it undermines rigid dichotomies between these categories and their supposed opposites: the global and the modern. Anthony Vital and Rita Barnard have argued that Mda's novel offers an unresolved contradiction between romantic idealist notions of place, indigeneity, and the natural (on the one hand) and (on the other) a postmodern skepticism regarding these categories. Up to a point, these readings are inarguable. Alongside its rigorous epistemological skepticism, the novel certainly has moments when it seems to succumb to the lure of valorized, stable conceptions of the traditional and the natural. Yet *The Heart of Redness* not only persistently brings into question clearly bounded idealist categories but also encourages critical distance from its own romantic lapses. Maintaining a skeptical posture, the novel suggests that any final representation of reality will be impossible: more narrative and dialogue will always be needed. The grounds offered for resistance and conservation constantly shift. The novel suggests that the local, the natural, and the indigenous must be seen as emerging and reemerging from specific, messy interrelationships with their supposed opposites. While *The Heart of Redness* very clearly represents the continued existence of local cultural and natural elements, it also notes how these have been transformed by "external" systems—colonialism, capitalism, the nation, and so on; yet even as they are translated, so do they translate those systems, making the latter heterogeneous. Thus, the local must be understood as a process or a set of processes.[16] Such a conception of place can be closely related to cosmopolitan bioregionalism, which both encourages attempts to understand local places and suggests that these places can only be grasped by engaging with the messiness and resistance to categorical closure entailed by their relationships with more global forces.

The Heart of Redness undoes easy categorization of identity and belonging, "unsettling notions of culture as stable and unified" (Vital, "Situating" 306) and demonstrating an "ambivalence about proper places and contexts" (Barnard 161).[17] The novel achieves this effect in part through its focus on the history of the village and its environs. If the text's historical memory creates a sense of place, of a rich and unique heritage accrued through time, it also emphasizes how this place has never had a singular identity and has always challenged efforts to create clear boundaries. For example, the novel stalks the category of the indigenous by emphasizing a precolonial history of conflict over land, displacement, and cultural interpenetration among the amaXhosa, the Khoikhoi, and the abaThwa, "those who were disparagingly called the San by the Khoikhoi" (73). Because of the long history of colonial and capitalist penetration, the present has resulted in an especially confusing cultural and political scene in which divisions based on the indigenous and the foreign, the local and the global, are impossible to maintain. Almost all traditional cultural practices and institutions have been transformed by their intersection with colonial modernity. New hexagons, with corrugated iron under the thatch, are now part of the local architectural landscape. Dances have been transformed as they become a means of making money and as the audience can "buy" a dance. In cases such as these, traditional institutions and practices continue to give a local shape to national and global systems, but they are inflected by those systems as well.

A particularly interesting example involves changes in the landscape and the village economy, resulting from the building of holiday "cottages" by people from "East London and other cities" (68). Upon learning of this development, Camagu is surprised, since "the land in the rural villages is not for sale. It is given by the chief and his land-allocations committees" (68). However, what he learns is that the significance of this traditional means of maintaining communal land has been changed; "the white people" and "some well-to-do blacks" bribe the chief. In the past they used "a bottle of brandy," but now they "bribe the chief with cellphones and satellite dishes" (68). If this situation suggests vestiges of traditional land tenure among the amaXhosa (which the colonists did their best to undermine following the cattle killing), it also points to their utter transformation by the combination of postapartheid tourism in the former homelands, vastly unequal economic relationships in South Africa, and political corruption.[18] In actuality, many holiday cottages have sprung up along the Wild Coast after their owners bribed traditional leaders who administer the land. Built in ecologically "sensitive and coastal conservation areas" without "any measure of control," these cottages have begun to have a significant negative environmental impact (Simukonda and Kraai 44–45).

Notions of purity are debunked in *The Heart of Redness* not only in respect to identity and culture but also in terms of the category of nature. At the most obvious level, the novel skewers the notion of a pure nature or wilderness that is not already intertwined with the social. In his landmark essay "The Trouble with Wilderness," William Cronon points out the problems with such a notion, delineating the ways that it occludes human history and even prevents a sense of ecological responsibility, since we are left with an either/or situation regarding impact—either there is no impact, or such impact is negative. The novel's representations of what Alfred Crosby calls "ecological imperialism" are interesting in this regard. In the present, the bush is covered with various kinds of nonnative plants that have been imported; some of these species—such as the wattle, the inkberry, and the lantana—present a serious danger to the continued existence of native species. The invasion and threatened conquest by this biota represent a kind of mixture of cultural and natural processes. Many of these plants have been intentionally imported, but their progress has not necessarily been anticipated or controlled by humans. As a result, the transformed landscape and methods of intervention cannot be understood in terms of a nature/culture dichotomy, and those who fall back on such simple categories appear misguided. We see these dynamics played out in the repeated criticisms of Qukezwa for her attempt to destroy the invasive species. Behind these criticisms is a notion of an unadulterated nature that she is wantonly attacking. When Qukezwa is brought to trial by the village elders for cutting down invasive species, even Camagu, before he understands her reasoning, thinks to himself: "What came over Qukezwa to make her chop down trees, when she has always presented herself as their protector. Part of her objection to the planned holiday paradise is that the natural beauty of Qolorha-by-Sea will be destroyed. But here she is, standing before the graybeards of the village, being charged with the serious crime of vandalizing trees" (213). For Camagu, all the trees are part of a singular nature that must be protected. During the trial, Qukezwa emphasizes that the wilderness has already been transformed by human history; as a result, conservation requires intervention in what appear to be natural processes.

Nonetheless, Qukezwa's conservation efforts raise some thorny issues regarding the novel's treatment of the categories of the indigenous and the natural. After all, she is trying to save "indigenous species" from the threat of the foreign. In addition, she seems to embody the ideal of an untutored indigenous knowledge of local ecology and of the proper relationship between humans and nature. Such suggestions in the novel lead critics to consider her character as Mda's means of maintaining idealist notions of the indigenous and the natural. Anthony Vital, for example, sees her as embodying a kind of radical ecofeminist ideal of women's closeness to nature. He argues that in *The Heart of Redness,* Qukezwa and more generally the female embody "a connection with nature the male feels he has lost," as well as "traditional visionary relations with nature," which are valorized by the novel (310–11). Such characterizations suggest that Mda reinforces the notion of an unchanged indigenous ecosystem that existed before the impact of colonialism and advocates for a return to an indigenous, properly ecological relationship between human and nonhuman nature—a natural society and culture. However, the novel suggests that traditional relations with nature have been neither singular nor untroubled and that there was not and cannot be an unchanging indigenous ecosystem untransformed by human activity.

While the novel advocates for the protection of indigenous species, it also suggests that the larger contexts of which these species are parts have substantially changed in such a way that there is no access to an underlying indigenous natural spirit that could be salvaged. The entire natural environment has been utterly transformed by the history of human and nonhuman invasion; to completely destroy "foreign" species—to strive for an ecological purity—is not a realistic course of action. On the whole, Qukezwa's environmental project is made problematic by its focus on such transformation.[19] Yet, to some degree, Qukezwa herself suggests the limitations of a project focused on ecological purity. At one point, she is asked, "Are you going to go out to the forest of Nogqoloza and destroy all the trees there just because they were imported from the land of the white man in the days of our father?" She responds, "The trees in Nogqoloza don't harm anybody, as long as they stay there" (216). Both Zim and Qukezwa note how the place that would seem to be the very epitome of an indigenous nature, Nongqawuse's pool, has been transformed. Zim tells his daughter, "In the days of Nongqawuse there were aloes [around the pool]. . . . Even when we were growing up, there were aloes. Also reeds used to cover this whole place. Only forty years ago . . . when I was a young man . . . there were reeds" (46). In *The Heart of Redness,* defining the natural identity of a place is complicated by ecosystemic changes over time, in part through the impact of "the migration of assorted flora and fauna" (Thomashow, "Toward" 129) The combination of attention to ecology and to the challenges it can pose for definitions of a "bioregion" affiliates the novel with a cosmopolitan bioregionalist sensibility.

In line with the novel's destabilizing of unified notions of culture and identity through historical narrative, *The Heart of Redness*'s representations of nineteenth-century amaXhosa society bring into question whether its relationship with nature was ever singular or perfectly harmonious. King Sarhili, we are told, "retreated to Manyube, a conservation area and nature reserve where people were not allowed to chop trees or hunt animals and birds. He had often told his people, 'One day these wonderful things of nature will get finished. Preserve them for future generations'" (131–32). These references to the drive for conservation among the Xhosa in the past suggest that even before the extensive spread of capitalism, any supposed balance of nature had been significantly impacted by what Sarhili saw as destructive environmental practice. Furthermore, the episode points to an existing conflict regarding such practice, a conflict possibly inflected by social status. (The king wants a conservation area in part because it allows him a place "to think things over in a peaceful environment" [132].) These cultural divisions continue to be reflected in the present in the debates regarding conservation. Despite the efforts by both sides to represent themselves as cultural preservationists, the issue is which thread of "tradition" will be followed. The believers accuse Dalton, the local trader (of "English stock" but with "an umXhosa heart" [8]), of trying to "change the ancient practices of the people" when he tries to prevent boys from "taking the eggs of birds from their nests" and "hunting wild animals with their dogs" (147). However, he counters with the story of Sarhili: "Perhaps you need to learn more about your forefathers. . . . King Sarhili himself was a very strong conservationist" (165). Given moments such as these, it is difficult to claim that the text supports the image of the eco-Indigene, in which indigenous peoples are envisioned as living in perfect harmony with nature and having an ideal ecological wisdom.

Admittedly, the question of epistemologies of nature is vexed in *The Heart of Redness*. Through the characters of the colonial governor Sir George Gray and the developers, Mda articulates a colonial nature/culture binary that constructs "nature" as wild and primitive, to be subdued and made useful through European rationality. In contrast, characters like Qukezwa and her father, Zim, seem to point to indigenous spiritual and aesthetic ways of knowing nature that emphasize interdependence, challenge the nature/culture binary, and can serve as a foundation for indigenous principles of conservation. However, the novel does not necessarily equate these ways of knowing with a *singu-*

lar traditional, indigenous relationship with nature in the present or the past. One thread of tradition may lay the groundwork for a locally rooted form of ecological care, but *The Heart of Redness* suggests there are other, contradictory environmental traditions. If the novel's focus on the history of a local community's relationships with nature aligns it with bioregionalism, the emphasis on the plural, even contradictory nature of these relationships disrupts any easy, singular conception of the bioregion's ecocultural identity. Similarly, Mda points to the existence in the West of a romantic reverential approach to nature in Camagu's recognition that a "green" backpacker hostel would have a strong appeal to certain kinds of tourists. ("There are many people out there who enjoy communing with unspoiled nature" [201].) Finally, the novel suggests that even those aspects of indigenous tradition that entail spiritual reverence for nature do not necessarily embody ideal ecological wisdom or practice. In a discussion of punishments for environmental transgressions, one elder notes that some tourists have been arrested (by the government) for smuggling cycads, while some village boys have been punished by village authorities "for killing the red winged starling." The other elders raise an outcry at even linking the two "crimes": "There can be no comparison here, the elders say all at once. The isomi is a holy bird. It is blessed. No one is allowed to kill it" (216). In this situation, animistic beliefs invest some species with spiritual significance while making others relatively unimportant (spiritually). Rather than being the grounds for care of biodiversity, these beliefs become a means of sanctioning biopiracy. Of course, one of the ways that *The Heart of Redness* opposes the terms of colonial discourse and the forms of exploitation it enables is precisely by challenging static, unitary categories such as "indigenous" (vs. "Western") and "nature" (vs. "culture").[20]

The most significant problem with the reading of Qukezwa as an image of the female eco-Indigene may be that it suggests *her* character is static and singular and that she represents the voice of truth in the novel. *The Heart of Redness* is far more resistant to closure than such a characterization would suggest. Qukezwa's knowledge and attitudes are not portrayed as only or even primarily the product of tradition or of nature; she understands not only the local ecosystem but also how tourism development schemes can work, and her desires are in part conditioned by the capitalist economy and the perception of geographic opportunity it encourages. (Near the opening of the novel, she tells her father than she has a "yearning" to go to Johannesburg because she will be able to "earn better money . . . I'll be somebody in the city" [46–47].) Moreover, both Qukezwa and Camagu grow in the novel precisely to the degree to which, in their dialogues with each other, they recognize the limitations to their own fixed categories and narratives of identity. If at first Camagu is condescending with Qukezwa, believing his education and experience give him a superior understanding of the world, he nonetheless comes to respect her knowledge and her perspective on development. Qukezwa also initially draws on her own clear dichotomies in judging Camagu, seeing him as representative of a modernity that is in strict opposition to traditional beliefs and practices; she believes his forms of knowledge are useless for Qolorha. Over the course of the novel, however, Qukezwa comes to realize not only that his cosmopolitan education and experience are not necessarily at odds with traditional practices but also that his knowledge can benefit the village as he pushes to get Nongqawuse's valley declared a heritage site by the government in order to block the casino scheme. Furthermore, just as she instructs Camagu about the operation of capitalist development, on which he is supposed to be the expert, so does he come to be able to correct her concerning local ecology. Near the end of the novel, when they are both watching a bird, Qukezwa claims it is a "hammerkop," but Camagu accurately points out that it is "the giant kingfisher" (221).

Thus far, my argument has emphasized the ways that *The Heart of Redness* debunks binaries and the static categories they entail. As a distinctly *postcolonial* text, the novel joins its deconstructive moves with a focus on ways in which such categories are connected to the legacy of imperialism. For example, Sir George Grey uses the civilized/barbarous dichotomy to sell a vision of advancement, but, in trade for "the greatest gift of all: education and British civilization," the amaXhosa must relinquish their land and political control (137). If this relationship between nineteenth-century imperial rhetoric of salvation and dispossession is fairly transparent, Mda more subtly suggests a connection between the cattle-killing movement's redemptive narrative and imperialist imperatives. In Mda's narrative, the believers' faith in Nongqawuse's prophecies, which offer the promise of transcending a messy and frightening reality through *external* agency, prevents them from grappling with their conditions of existence themselves. Not surprisingly, their belief makes them "lethargic" and leads to their eventual starvation. Sir

George Gray is then able to succeed "beyond his wildest dreams in turn-ing [the Xhosa] into 'useful servants, consumers of our goods, contrib-utors to our revenue'" (Peires 321).

In the present, the locals once again threaten to facilitate their own dispossession and loss of agency by buying into a narrative of salvation—now embedded in the discourse of economic development—based on the old colonial binaries. This danger is especially apparent in the argu-ments regarding the proposed plans for the casino complex. Not only do Bhonco and his band of Unbelievers accept the story of economic devel-opment for the community (i.e., that the scheme will bring jobs and fur-ther social infrastructure), but they also claim that it will offer deliver-ance from all that is primitive and wild: "We want to get rid of this bush which is a sign of our uncivilized state. We want developers to come and build the gambling city that will bring money to this community. That will bring modernity to our lives, and will rid us of our redness" (92). Ironically, the new development will be partly advertised by holding out the promise of the wildness of the Wild Coast, in particular for surfers: "the waves here are more suited to the sport than the waves of other big cities in South Africa. The waves here are big and wild" (199). In this case, development feeds off its apparent opposite, the "wild," suggesting that these categories are the twin poles of the same construction. The creation of discrete categories makes ideas of wilderness and untamed nature easily commodified and enables them to be brought more easily into a discourse of modernity.

The novel's suspicion regarding rigid, hierarchal models of identity and culture is also reflected in the debates between Camagu and Dalton regarding effective means of resistance to the casino. Though Dalton is among the leaders of the resistance, Camagu warns him that his efforts to help the people are bound to fail unless he is willing to relinquish au-thoritative (and authoritarian) representation; to truly empower them, he must become an active listener: "You know that you are 'right' and you want to impose those 'correct' idea on the populace from above. I am suggesting that you try involving the people in decision-making rather than making decisions for them" (180). Camagu seems to be sug-gesting the need for Fanonian intellectuals who will develop represen-tations of the people, their needs, and possible courses of action in dia-logue with them. In "On National Culture," Fanon insists that the native intellectual "must join [the people] in that fluctuating movement which they are just giving a shape to, and which, as soon as it has started, will be the signal for everything to be called in question. . . . It is to this zone of occult instability where the people dwell that we must come" (227). If the call to join "the people" urges the intellectual to identify with a subaltern collective, the references to the "fluctuating movement" and "zone of occult instability" in which "the people" are to be found suggest that the identity of this collective is always in progress. In fact, a primary theme running throughout "On National Culture" is a warning against seeking liberation through a static collective identity. The Fanonian aspect of Camagu's appeal becomes especially apparent when he questions Dalton's "cultural village," which produces an exoticized version of traditional culture. As Camagu notes, "I am interested in the culture of the amaXhosa as they live it today, not yesterday. The amaXhosa people are not a museum piece. Like all cultures, their culture is dynamic" (248).

Camagu's perspective on the best means to protect Qolorha is expressed not just in his discussions with Dalton but also, more positively, in his own efforts to counter the casino scheme and to develop an alternative in a community-based form of ecotourism. The novel mostly endorses Camagu's efforts, and in this sense his Fanonism points toward the (shifting) grounding offered by The Heart of Redness for successful protection of local peoples, cultures, and natural environments in the face of contemporary imperial threats. When asked by the developers, "If you fight against these wonderful developments [to be brought by the casino], what do you have to offer in their place?," Camagu answers, "The promotion of the kind of tourism that will benefit the people, that will not destroy indigenous forests, that will not bring hordes of people who will pollute the rivers and drive away the birds." He then proclaims, "We can work it out, people of Qolorha" (201). As this last statement indicates, Camagu's idea, which develops into the ecofriendly backpacker hostel, avoids the pitfalls of Dalton's project through its focus on collective participation; the hostel is planned, built, and owned by the village.

If there are echoes of "On National Culture" in Camagu's approach to antiimperial forms of collective identity and agency, however, The Heart of Redness also significantly departs from Fanon's vision. Most obviously, Mda is not focused on the category of the nation; being both before and after the nation, Mda's local community outstrips the abstraction and teleology that the category entails. Even more important for my argument here, unlike Fanon, Mda represents the search for community in terms of a wider ecosystem. When Camagu asks Qukezwa how she knows so much "about birds and plants," she responds, "I live with them" (105). In the context of the novel, this "living" with nonhuman nature is not just a matter of living *in* a particular environment but also of seeing oneself as part of a community. The focus on ecological affiliation and on such affiliation as one basis for geopolitical identification is perhaps the most salient bioregionalist feature of *The Heart of Redness.*

The notion that humans are interdependently related to wider ecological communities comes into particular relief in the novel's representations of the ways that abrupt ecosystemic changes have drastic implications for human societies and cultures. Lung-sickness is a particularly good example. It was caused by microbes that came from Europe and had a dramatic impact on amaXhosa society and culture. The novel stresses that the disease cannot be reduced to social or cultural explanations, although it is tightly bound with social processes both causally and in terms of its impact. Its appearance represents something new, which the amaXhosa must try to understand in all its complexity without relying on preconceptions and definitive categories. In Mda's narrative, one problem with the cattle-killing movement is that it relied on an existing grand narrative to explain a transformation in the natural environment. Like "the people," nature must be listened to and actively engaged with as it emerges in the present.

As Mda's version of the cattle killing suggests, in *The Heart of Redness* a community must try to understand the wider system of ecological relationships within which it finds itself if it will determine effective courses of action for its survival. The novel makes it clear that the destruction of both resources and Qolorha's unique beauty in the wake of the casino's construction will have a dramatic impact on the community's situation; they will very quickly become completely transformed and have little control over their future. By understanding the relationships in which they find themselves, and grasping the significance of those relationships, they have a better sense of how to respond to the casino scheme. The same could be said about the issue of ecological imperialism. The community needs to understand the particular threat they face by closely examining the set of relationships in which it is embedded. At the end, through a better (if necessarily incomplete) understanding of these relationships and their significance, the community has been able to fight off both the casino scheme and the threat represented by invasive species. Driving back to Qolorha, Camagu "sees wattle trees along the road. Qukezwa taught him that these are enemy trees. All along the way he cannot see any of the indigenous trees that grow in abundance at Qolorha. Just the wattle and other imported trees. He feels fortunate that he lives in Qolorha. Those who want to preserve indigenous plants and birds have won the day" (277). Such preservation is important, in part, because of the issue of local control over resources, one of which is, precisely, the place's uniqueness; for, of course, uniqueness is a commodity that draws people to Qolorha and its backpacker hostel. The casino scheme, by contrast, would render Qolorha a copy, something that could be on any coast; the developers, savvy in the production of such projects, would be the ones in control.

This reading links *The Heart of Redness* with a bioregional perspective. The novel focuses on the value of reinhabitation, the "commitment to understand local ecology and human relationships" (Thomashow, "Toward" 125), and, in particular, on the ways that understanding and appreciating the uniqueness of a particular place, defined culturally and ecologically, can contribute to resistance against colonial modernity. Mda may emphasize geographical and categorical intermixing, but he does not suggest that differences among places have no meaning. In fact, in *The Heart of Redness* effective action is dependent on recognizing and bringing together the knowledge and perspectives offered by such differences. As Camagu lives in Qolorha and engages with Qukezwa, the local comes to represent not only the gaps in his education but also something to be known—no matter how incomplete that knowledge will of necessity be. At the same time, as Camagu becomes more and more involved in the resistance movement against the gambling complex development scheme, his knowledge from Johannesburg and the United States becomes crucial in the success of the movement. The high degree of interpenetration between categories (urban and rural, local and global, regional and national) gives the differences in these categories more not less meaning.

Nonetheless, these differences are not to be understood as ahistorical or clearly bounded. Opposing the casino scheme and determining the best alternative—identifying what and who needs to be opposed, as well as why—requires thinking through the interplay of similarity and difference in shifting terms based on context and scale. It is important, for example, for Camagu to keep in mind both how Qolorha is linked with national majorities in terms of dispossession by elites *and* the differences that result from Qolorha being a marginal, rural place in respect to the urban centers of South Africa. Similarly, the novel suggests how Dalton is both different from some of his liberal white friends who have little sense of identification with places like Qolorha (or even South Africa as whole) and, at the same time, how his interests are not necessarily identical with the interests of the majority of the villagers.

It should be noted that Camagu's solutions are neither ideologically pure nor necessarily adequate. He maintains that he opposes the exoticization and commodification of Xhosa culture embraced by Dalton's cultural village, but his own cooperative society produces "traditional isiXhosa costumes and accessories . . . to be marketed in Johannesburg" (161). In addition, the backpacker hostel is sold through the claim that Qolorha can offer "unspoiled nature" and, as result, draws on a commodified notion of wilderness challenged by the novel as a whole. Most important, at the close of the novel, Camagu notes, "Sooner or later the powers that be may decide, in the name of the people, that it is good for the people to have a gambling complex at Qolorha-by-Sea. And the gambling complex shall come into being" (277). This final comment suggests that there are limitations to Camagu's strategies for resistance. In his focus on local solutions, he has not taken into consideration fully enough the protean operation of imperial capital, working at both national and international scales. As a result, his strategies may very well be only temporarily successful in terms of conservation of local control and environment. However, *The Heart of Redness* also suggests that it is impossible to imagine the means of completely transforming the current set of unequal political and economic relationships working at both the national and global levels, especially since one must always address how they manifest themselves locally and are inflected by local processes. The necessity of working through the messy dialectical interrelationships between local place (the particular) and more abstract geographical entities (the general) means that any solution will always be provisional, bound by time and place, and in need of revision. There is no once and for all. In this sense, the novel not only makes no claims to the authority of particular representations and solutions (including its own) but also encourages us to seek out their limitations. Ironically, in so doing, we end up endorsing the one truth it does insist on: the impossibility of a "destination beyond which all knowledge ends" (97).

As the foregoing analysis implies, if one is to align *The Heart of Redness* with bioregionalism, that term must be qualified with "cosmopolitan." It is worth recalling that a *cosmopolitan* bioregionalist sensibility is "open-ended," perceiving boundaries as "permeable" and identities as "pluralistic" (Thomashow, "Toward" 129–30). At the same time, the term *bioregional* denotes a difference from a cosmopolitan approach to place more generally.[21] The latter will not *necessarily* take account of the importance of ecology or the significance of ecological relationships for humans. To discuss place in *The Heart of Redness* necessitates that one appreciates the importance of how Qolorha has been shaped by particular relationships between nature and culture, as well as by relationships between the local and global.[22]

The Heart of Redness is also closely aligned with an environmental justice perspective. It may support the protection of biodiversity, but it also emphasizes how the significance of conservation is tied to its impact on livelihoods and economic well-being of local communities. More generally, in this novel debates about conservation—for example, about what should be conserved and why—are always closely connected with struggles over power and economic resources and are mediated by gender, economic status, age, race, history, geography, and culture. Furthermore, the text strongly emphasizes the need to take into account local knowledge and perspectives if conservation and development will be both effective and socially just. Unlike *Cry, the Beloved Country* and *When Rain Clouds Gather, The Heart of Redness* strives to imagine ways to combat imperialism's destructive legacies and projects that will not generate reformulated colonial assumptions and relationships.

In particular, Mda disrupts the roles of teacher and leader associated with colonial conservation. It is true that Camagu comes up with the general ideas that lead to the defeat of the casino project, at least temporarily, and to the alternative ecotourism development project. However, through much of the narrative, he needs guidance from members of the community not only about the local environment and development but also to gain a more informed perspective on himself and the world. Even toward the end, Camagu does not exactly become a leader and is certainly not a savior figure. Dalton is the one who tries to inhabit those roles, driving up at the last minute with the document designating Qolorha as a national heritage site and putting a stop to the casino. Just as important, the text emphasizes the uncertainty of the meaning of the changes Camagu has helped bring about. Is Qolorha saved? What are the long-term limitations to Camagu's ideas? As is true throughout *The*

Heart of Redness, shifts in scale, both temporal and geographic, render meaning uncertain. Thus, any authority and position Camagu might be ascribed in the community are necessarily unstable.

Mda's cosmopolitan bioregionalist sensibility is a crucial component of its relative (but not complete) avoidance of the discourse underpinning colonial conservation, as well as of its alignment with environmental justice. It projects identity as heterogeneous and decentered as a result of the transformative intersection of inside and outside, history and geography. Such representation challenges claims to objective knowledge, as well as unified and noncontingent categories (nature, race, and place, for example) that could ground a "natural" authority. *The Heart of Redness* encourages both skepticism toward the authority of any single representation or voice and the effort to take into account the multiple, heterogeneous perspectives in place (and beyond) to formulate an always uncertain trajectory for justice, sustainability, and conservation.

Since the publication of *The Heart of Redness,* two new threats have changed the landscape of environmental justice on the Wild Coast. First, an Australian company, Mineral Resource Commodities Limited (MRC), has been pushing for over a decade to extract heavy minerals from the coastal sand dunes around the Xolobeni community in Pondoland. The targeted land is communal but managed by traditional leaders and local government. Second, the government wants to replace the existing N2 road with a toll highway to provide a shorter, more efficient route through the Wild Coast, a project that many claim is driven by mining interests (Simukonda and Kraai; Dellier and Guyot).

Although these threats are clearly different from the casino project in *The Heart of Redness,* the situations have striking similarities and raise some similar issues. The Maputaland-Pondoland region is renowned for its plant diversity, endemism, natural beauty, and of course "wildness." As a result, some tourism interests and external environmental groups have fought the projects. Local community opposition is primarily focused on livelihoods, health, and loss of land. Many in the communities claim that they have not been adequately consulted or informed. Meanwhile, the mining operation and the road have some local support based on the promise of jobs and economic development. This support is strongest among community leaders and local politicians who would most benefit. The situation has resulted in significant communal conflict; "neighbors who once lived in peace and harmony" no longer speak "because they belong to one of the two opposing factions—for or against the mining project" (Simukonda and Kraai 47). The national government has pushed for both projects, claiming that they would be in the national interest and would create economic growth for this historically disadvantaged area. However, most of the benefits would not accrue to the Wild Coast, and the majority of the profit from mining would not even stay in South Africa. Nevertheless, because of the powerful interests involved, the Xolobeni mining project "may soon become a reality, unless UNESCO is willing to proclaim the Xolobeni area a World Heritage Site" (Dellier and Guyot 95).

Even more than *The Heart of Redness,* this situation highlights the importance of thinking in terms of complex, unequal relationships operating across geographical scales. It also reinforces Camagu's worries about the kind of solution embraced by Qolorha to destructive development on the Wild Coast. In the pursuit of their interests, foreign and national elites come up with new strategies, new venues, or new kinds of projects; meanwhile, vast economic and political inequalities continue to entrench procedural injustice. In response, opponents must be vigilant and flexible, able to maintain a militant particularism while reaching across boundaries. In the fight against the mining enterprise and the toll road, various kinds of actors have joined forces—wilderness advocates with a deep green hue, those working within the tourist industry, regular members of local communities, and some traditional leaders. These groups have found points of connection, even as there remain tensions among them regarding issues of land ownership and resource access. This kind of coalition building and, more generally, the struggle against environmentally disastrous forms of development require adaptability born from an awareness of alternative perspectives shaped by a history of vast inequality. This is a point also made by Nadine Gordimer's novel *Get a Life,* which evokes a number of recent environmental conflicts in South Africa—including the struggle against the Xolobeni mining project and the N2 toll road.

The Place of Conservation: Nadine Gordimer

To some degree the difference between environmental justice and other kinds of environmentalism can be defined in terms of the kind of skepticism evinced by Mda regarding naturalized external expertise, technical solutions, and the separation of categories on which they depend. Yet *The Heart of Redness* does not represent a point along some teleology of environmental justice. In fact, environmental justice movements often resist any universalist or global narrative of development; they suggest, instead, that the means and ends of resistance will emerge in the context of specific historically and geographically situated conflicts. David Schlosberg notes the dangers in seeking "uniformity" in environmental justice activism; it can "limit the diversity of stories of injustice, the multiple forms it takes, and the variety of solutions it calls for" (535). While skepticism regarding (for example) the separation of environmental problems and political relationships is a crucial part of any environmental justice theoretical framework, such a framework also necessitates reading meaning in ways that resist linear assimilationist analysis. The significance of this point in terms of South Africa's recent history can be extrapolated through a contrapuntal analysis of Nadine Gordimer's two novels *The Conservationist* and *Get a Life*.

Published three decades apart, both of these novels can be broadly associated with environmental justice in foregrounding the contradictions inherent in colonial conservation and their emphasis on the connection between any environmental project and questions of social power and privilege. However, ultimately, the two novels deploy the figure of a white (male) identifying himself as a "conservationist" very differently. The earlier novel's focus on antiapartheid struggle and the connection between colonialism and ecological management results in the erasure of that figure from the vision of a better future. In contrast, *Get a Life* is more uncertain about the grounding for anticolonial struggle and positions the conservationist (in this case an ecologist) as potentially part of rather than necessarily outside the struggle for social justice in South Africa. This difference between the two novels need not be framed using a narrative of progress (or decline) in Gordimer's understanding of imperialism as neoliberalism, of socioecological injustice, or of the significance of ecology, but can be framed in terms of different historical moments encouraging divergent visions of the path to a more just future. The truth or error of these visions always remains uncertain, as the ambiguous present necessitates a flexible approach to temporal narratives and challenges a teleological disciplining of meaning.

Lawrence Buell's concept of the environmental unconscious offers a useful means for exploring the relationship between Gordimer's two novels and the resistance of environmental justice frameworks to the separation of politics and environmentalism. According to Buell, the environmental unconscious is the result of "habitually foreshortened environmental perception" (*Writing* 18). It refers to all those aspects of the physical environment that are suppressed in our awareness by inattention, ignorance, specialized training, conventions of language, and so on. According to Buell, the environmental unconscious has two aspects. Its negative aspect "refers to the impossibility of individual or collective perception coming to full consciousness at whatever level: observation, thought, articulation, and so forth" (22). Its positive attribute is to be found in the potential for breakthrough achieved in the bringing to awareness and then to articulation of what has been suppressed. Through this second attribute, the environmental unconscious can become "an enabling ground condition" for a productive reenvisioning that enables "a fuller environmental(ist) sense of [site] than workaday perception permits" (23). Finally, the notion of the environmental unconscious can encourage a productive resistance to closure in environmental representation since it suggests that such representation will always necessarily entail suppression of aspects of environment.

Yet Buell's own formulation of the environmental unconscious can serve as evidence not only that all acts of environmental perception and representation involve repression in their inevitable forms of closure but also of how difficult it is to resist such closure. Buell acknowledges that he has drawn on Frederic Jameson's concept of the political unconscious, but he also separates Jameson's concept from the "environmental unconscious" by privileging environment as a determinate of identity. For Jameson, individual and collective identity—and, as a result, all literary texts—are mediated by ideological configurations and the political relationships that generate them. Every literary text is a rewriting of a prior ideological subtext. Contrasting himself with Jameson, Buell argues that "embeddedness in spatio-physical context is even more intractably constitutive of personal and social identity, and of the ways that texts get structured, than ideology is, and very likely as primordial as unconscious psychic activity itself" (24). As Buell's reference to the "primordial" indicates, he seeks an uncovering of the ways that environment has shaped us and our understanding of the world that is more fundamental than the constitutive function of forms of consciousness determined by social roles and political relationships. This position is incompatible with Nadine Gordimer's perspective, even though her writing is profoundly engaged with what I call the socioecological unconscious. For Gordimer, the political fundamentally determines identity and reality, including the very shape of what is supposedly beyond its influence. This is not to say that she downplays the role of "embeddedness in spatio-physical context." Ideology and environment are involved in a mutual determination in her novels. We see this, in particular, in the way that she constructs the relationship between the political and the natural. There is no nature and no concept of it that can be separated from the shaping influence of ideology; at the same time, ideology can never escape the impact of the natural. Ultimately, for Gordimer, gaining insight from the realm of "the environmental unconscious" entails a transgression of existing distinctions between nature and politics, the environment and ideology. For Gordimer, especially in *Get a Life*, environment resists rather than reinscribes analytic closure.

The Conservationist offers numerous examples of what might be characterized as the operation of the socioecological unconscious. The novel is focused on the processes of ego-formation (and maintenance) under apartheid and on the forms of repression that these processes entail. It is, in particular, focused on the subjectivity of a rich white industrialist who owns a farm just outside of Johannesburg. In the novel, his hegemonic perspective is often brought into question through glimpses of what he cannot entirely suppress but also what he cannot acknowledge without a substantial shift in subjectivity. These glimpses frequently entail aspects of the "spatio-physical" environment. The novel insists that the development of subjectivity must be understood in terms of the distribution of land and possessions determined by systemic relationships, especially those entailed by apartheid capitalism. Yet the material world is also precisely where the repressed lies; that which holds in place Mehring's ego is where the psychic waste threatens to come to the surface.

Mehring's farm is especially important in terms of understanding how the organization of land and things maintains his identity. He conceives of the farm as a spot of idyllic, managed nature separated from the dominant economic and political system through which he has made his money. For him, it naturalizes his sense of autonomy, belonging, and uniqueness: "a sign of having remained fully human and capable of enjoying the simple things of life that poorer men can no longer afford" and of a certain freedom, "not the freedom associated with a great plane by those who long to travel, but the freedom of being down there on the earth, out in the fresh air of this place-to-get-away-to" (22–23). Just as he separates the farm from the system, he believes he too escapes control. As his mistress says to him, "Ah yes, that's the trouble—you think you are inviolate" (107).

The farm is also evidence of his knowledge of and ability to manage nature: "Reasonable productivity prevailed; he had to keep half an eye (all he could spare) on everything, all the time, to achieve even that much, and of course he had made it his business to pick up a working knowledge of husbandry, animal and crop" (23). He envisions himself as the one best able to constructively regulate nature in a way that those who live on the farm cannot. He is both the bringer of order and the protector of its natural beauty and purity. Through these roles, his natural right to ownership is secured. He is the land's proper steward. At the same time, through his management of nature, he reveals that he rises above and is autonomous from it as well. Ultimately, his wealth and power are only signs of such unique abilities and identity.

One of the deep ironies of the novel, however, is that the farm itself can offer glimpses of a very different image of Mehring than the one he normally sees reflected back. In contrast with his perspective, the novel itself emphasizes how his identity and the function of the farm in relation to it are produced by apartheid capitalism. For example, his ownership of the farm enables him to repress precisely how his acquisition of it results from dispossession and exploitation, which have also produced the very environmental threats that he supposedly struggles against in his role as a "conservationist." As he himself acknowledges, "To keep anything the way you like it for yourself you have to have the stomach to ignore—dead and hidden—whatever intrudes" (79). What Mehring ultimately strives to conserve is not the natural integrity of the farm, but an organization and image of it that allow for his pleasure and reinforce his identity. The novel's antihegemonic vision is brought into focus when that which is repressed surfaces, often through Mehring's consciousness of events. Most prominent here would be the material traces of the history of dispossession and its effects, traces that he perceives as forms of environmental threat: poaching, trespassing, litter, and so on. The dead and only superficially buried body is both an example of theses traces and the symbol of them. The body is of a "city slicker" who has apparently been killed in some kind of township violence and then dumped on the farm. In Mehring's view, this body represents an incursion on his land—both trespassing and littering; it is a form of pollution, and as such, Mehring wants it removed as quickly as possible. But the body is not removed; instead, the police bury it sloppily, as becomes apparent later on when it resurfaces after the storm. In the course of the novel, the body becomes an image both of Mehring's lack of mastery and of how the violence and waste produced by the system that has given Mehring his wealth are an integral if buried part of the farm. *The Conservationist* is focused on questioning the boundaries—both geographical and categorical—that maintain the hegemonic order and the forms of identity it entails. In particular, the novel undoes what Anthony Vital refers to as "the sheltering distinction between inside and outside, between order and litter," through which colonial modernity is distanced from the waste it creates and through which its beneficiaries are able to uphold the value of themselves and their places in opposition to the people and places that have been wasted ("Waste" 191). In this sense, the body as it becomes part of the farm reflects back to Mehring precisely a part of himself that he does not want to acknowledge and that threatens his identity. The novel traces the oscillation of a coming to light of such aspects of Mehring's material environment and his efforts to suppress their disruptive potential.

Mehring's notions of nature are an important part of the epistemological ordering brought into question in the novel. In his pastoral vision, it is a realm of eternal value and beauty, clearly bounded from social processes and from history. It must be carefully protected from black others who do not understand or appreciate it and who cannot make it productive. In the opening of the novel, Mehring comes across "pale freckled eggs" in the possession of some of the children on the farm (9). He conceives of their play as a transgression that endangers the guinea fowl on the farm and, more generally, as an example of the practices of the farm workers that threaten its beauty and value: "already the farmer has had occasion to complain about the number of dogs they are harbouring (a danger to the game birds)" (12). Ultimately, the children's game is an image of the threat to the earth represented by such others: "A whole clutch of guinea fowl eggs. Eleven. Soon there will be nothing left. In the country. The continent. The oceans, the sky" (11). We see this sense of threat, as well as the role it gives him as "conservationist," repeated numerous times: "he never leaves so much as a cigarette butt lying about to deface the farm; it's they—up at the compound—who discard plastic bags and put tins beside tree-stumps. He's forever cleaning up after them" (43). Mehring's concern reaches a kind of climax after a fire sweeps through his farm. He conceives of this fire as an invader that has come onto his land from elsewhere and ravaged his beautiful place: "The fire's territory: the invasion marked out with its inlets, promontories, and beach-heads. Taken overnight" (94). He believes his farm has been permanently damaged, made ugly and less valuable. He also believes the fire could only have come from the carelessness and ignorance of black others who do not understand and cannot care for nature:

> Will the willows ever be the same again? They think if the lands are saved no damage has been done. They don't understand what the vlei is, the way the vast sponge of earth held in place by the reeds in turn holds the run-off when the rains come, the way the reeds filter, shelter. . . . What about the birds? Weavers? Bishop birds? Snipe? Piebald kingfisher that he sometimes sees? The duck? The guinea fowl next in the drier sections, as well. . . . But what else—insects, larvae, the hidden mesh in there of low forms that net life, beginning small as amoeba, as the dying, rotting beginning again? (97)

Mehring believes he knows the operation of ecology; through this knowledge, he is able to understand the threats to the natural world and to protect it. In turn, the black majority's right to ownership of the land is denied, and their expulsion to places of "waste"—townships and homelands—is validated. They are the source of an environmental threat that must be contained. This perspective is the ultimate manifestation of what David McDermott Hughes describes as European settlers' efforts to legitimate their rule in Africa and to exclude Africans from power, wealth, and territory by establishing their own capacity to understand and represent the land.

Gordimer has herself noted that colonial conservationist thinking was one inspiration for her writing of the novel:

> As time went by, I found how it's such a paradox really because we're all for conservation; we all have this concern about the natural environment in which we live. But in the South African context, it often becomes something unpleasant and almost evil, as it did in *The Conservationist,* because there's the question of whose land? Can you own the land with a piece of paper, a deed of sale? So the concern for the birds and the beasts and the lack of concern for the human beings become another issue. (Bazin and Seymour 286–87)

As this comment suggests, the novel's focus on Mehring's conception of conservation and the notions of nature it entails makes *The Conservationist* useful in developing a postcolonial critique of certain kinds of environmentalist ideology. The novel primarily achieves this effect through its bringing to light of an environmental unconscious that points to all that Mehring's vision represses. Despite Mehring's belief in his own cleanliness, he is, in fact, a source of what he sees as "trash" on the farm, and this "trash" is picked up by the farm workers precisely because they do not see it as such. As Jacobus cleans out the ash trays in Mehring's farm house, he thinks to himself: "The butts were all smoked down to precisely the same length—like the ones the children knew they must deliver to him whenever they found them in the grass" (65). The guinea fowls are not in danger from either the children or the dogs of the farm workers; Jacobus, the leader of these workers, even tells Mehring "that there were plenty of guinea fowl about" (33). Mehring, however, will not listen. Later, he does see the birds, although he does not acknowledge his earlier mistake: "over there, over there, are twenty-three guinea fowl" (108). Similarly, the fire does not actually represent a threat to the ecosystem of which the farm is a part; it recovers, suggesting that what he saw as ruin and waste is actually part of the very landscape he deems the most valuable: "things come to life under his eyes as the syntax of a foreign language suddenly begins to yield meaning" (133). In these last two instances, Mehring does see what he did not before; in a sense, his ability to decipher the "foreign language" of the natural environment does advance.

However, in more substantial ways, Mehring's vision of his farm does not really change. He may see the guinea fowl and the recovery of the farm's flora and fauna after the fire, but he does not consider that they point to both the farm workers' environmental knowledge and the possibility that their practices do not represent the threat he considers them. He does not reflect on what the fire and its aftereffects might suggest about the development of the farm: that it needs neither him nor its boundaries for its protection. He also does not reevaluate what he considers waste, destruction, and decay, as opposed to conservation and growth. Finally, he does not rethink his conception of the relationship between natural processes and social processes. He remains as he is described at one point, "inattentive to the earth" (46).

In *The Conservationist,* this inattentiveness is necessary for Mehring because "the earth" challenges the strict forms of delimitation that he imposes on it and that are necessary for the maintenance of his authority and identity. In Gordimer's novel, it is the sublime, outstripping existing order and meaning. This notion is most fully figured in the storms that come in from the Mozambique Channel. These storms are awesome not only in appearance but also in their effects. They wash away roads and disrupt telephone communication, thus preventing Mehring from getting to and communicating with the farm. They also utterly transform the land itself: "The sense of perspective was changed as out on an ocean where, by the very qualification of their designation, no landmarks are recognizable" (233). The storms initially disorient Mehring because they exceed his knowledge of nature and his ability to master it. After a small culvert that he has frequently crossed overflows and finally sweeps an Afrikaner couple away to their death, Mehring is overwhelmed by the force involved, as well as by the thought that it could have been him held helpless and killed by the stream of water. However, by imagining a conversation with a secretary about the incident (in other words, by placing it in the context of his usual roles), he recontains the threat by suggesting that he would have known better than to have crossed the gully: "But I'd never do a thing like that" (237).

He attempts this same kind of recontainment when he gets back to the farm. Because of Mehring's forced absence, Jacobus has had to take over the running of the farm entirely. As a result, he has broken from his typical role in relation to Mehring, and this disruption has the potential to challenge Mehring's sense of his own position as rightful steward. At the same time, the storm has utterly transformed the farm's landscape, in part because waste that has been buried or hidden comes to the surface. As was the case with the incident at the culvert, this physical transformation potentially points to Mehring's lack of mastery over nature. The surfacing of the man-made trash also suggests the connection between the farm and those places of "waste" from which, in Mehring's view, it is separated and must be protected. In the face of the changes, Mehring again tries to reassert control upon his return by giving orders and considering the ways that he will restore the farm's proper form. He thinks both that "normal procedures must be returned to" and that he must "drain the land" so that it will not be turned to "swamp" (244–45).

However, the effects of the storm cannot be so easily contained. When he is left alone amid his flooded farm, he is overwhelmed: "for the first time since the flood, he is exposed to the place, alone: it comes to him . . . in its living presence" (245). This "living presence" both smells different ("a smell of rot") and looks different: "Something heavy has dragged itself over the whole place, flattening and swirling everything." These transformations, he is aware, have been made "by an extraordinary force that has rearranged a landscape as a petrified wake." As elements of Mehring's "environmental unconscious" are brought to the surface, he becomes aware of a natural system that shapes his farm and that he cannot understand or master. In the face of this knowledge, he "feels an urge to clean up, nevertheless, although this stuff is organic; to go round collecting, as he does bits of paper or the plastic bottles they leave lying about" (246). What he has been conserving is not nature or the environment, but his idea of them and his relationship with them. Faced with an aspect of the environment he has never admitted, he wants to clean it up in order to be returned to himself.

Immediately afterward, Jacobus finds the body that the storm has uncovered, and Mehring jumps in his car to escape. His strong reaction suggests that more than the body has been brought to the surface: "Recognized by the shoes and apparently what's left of a face, with the—that's enough! Why hear any more, it's not going to do anybody any good. That's enough. A hundred-and-fifty thousand of them practically on the doorstep" (249). The body points to the connection between Mehring's farm and the places of waste from which he wants to separate it; the former is shaped by and feeds off the latter. He has seen himself as transcending waste and its sources, but he and his farm are part of the system that creates them. There is no escaping this relationship: "the only thing that is final is that he's [the body's] always there" (251).

As Mehring escapes from his farm in his car, he picks up a female hitchhiker who takes him to an old mine dump with the apparent intention of seducing him. In this space, Mehring once again confronts the surfacing of aspects of his environment he has suppressed. The divisions that maintain his sense of the world and identity collapse, and this time he cannot escape. Shaded by a grove of eucalyptus trees, quiet, and separated from the highway, the dump itself initially appears as a pastoral spot. However, Mehring quickly becomes aware that it is a place covered in trash, and this brings home the fact that it is a place built from waste. He believed he was running from one such place, his farm, an apparent retreat that is actually a manifestation of the wasteful system; however, there is no escape—he and all his places have been shaped by that system. This inability to escape is also manifested in his interactions with the woman herself. She is able to manipulate him through his body, which is another aspect of a nature he does not understand or control, and his awareness of being watched by a man leads him to believe he has been trapped. Faced with the sense that he is in the grip of powers (social and natural) beyond him, Mehring has a mental breakdown.

In the final scene of the novel, the reader is returned to the scene of the farm, where the workers are holding a funeral for the dead body. The scene can be read as a representation of a potentially revolutionary dispensation in which the community has moved beyond the hegemonic mapping of identity by apartheid (as revealed by their sense of solidarity with the dead man, whom they initially avoided). We do not know Mehring's final fate, but this scene suggests his irrelevance in such a dispensation. It is the text's last incursion into the environmental unconscious—this time, revealing a potentiality that lies in the present. The scene (and the novel as a whole) has been criticized for reinscribing a pastoral sensibility (Gorak). However, as Rita Barnard points out, the acceptance of the dead man (a "city slicker") as "one of them" complicates this reading, since it "undoes the opposition between country and city" (90–91).[23] The scene is not a retreat from history; rather, its revolutionary potential includes its expansive representation of an environment that defies efforts to strictly separate categories (urban/rural, nature/history, environment/politics).[24]

In many ways, then, *The Conservationist* focuses on "the environmental unconscious" by bringing attention to aspects of environment unrecognized by characters and, possibly, by readers, as well as by drawing attention to the forces that create that unconscious. It also points to the transformational potential in the coming to awareness of suppressed aspects of the environment. Finally, the novel might be useful in bringing together the notion of "the environmental unconscious" and "the political unconscious." It points to the ways that environment cannot escape from being shaped by ideology, as well as the ways that environment is crucial for the operation and disruption of ideology. Yet, ultimately, in *The Conservationist* environment is secondary to ideology as a determination of consciousness, and ecology is secondary (at best) to politics as a concern. (In this sense, the novel reverses—rather than collapses—Buell's causal priority.) Environment only moves beyond this secondary status when it takes the form of nature as the sublime, in excess of human thought and in no need of protection.

In a more recent novel, *Get a Life,* Gordimer again can be linked to the notion of a socioecological unconscious in the ways she brings attention to the limitations of a prominent form of conservation. However, this time, ecology and environmentalist concerns are treated more ambivalently. If they remain constrained by their connection with bourgeois subjectivity, they also serve more socially positive (possibility even transformative) functions than they do in the earlier novel. In many ways, the sensibility of the novel takes its tone and perspective from the central character, Paul Bannerman, "an ecologist qualified academically at universities and institutions in the USA, England, and by experience in the forests, deserts, and savannahs of West Africa and South America" (6). Through Paul's perspective, the novel emphasizes forms of ecological connection and implications of human behavior that normally remain hidden and that are brought to light by the study of ecology; his training enables ways of seeing that break with the ordinary. "The work" that he and his colleagues do "informs their understanding of the world and their place as agents within it, from the perspective that everyone . . . acts upon the world in some way. Spray a weed-killer on this lawn and the Hoopoe delicately thrusting the tailor's needle of its beak, after insects in the grass, imbibes poison" (83). At the same time, Paul becomes aware in the course of the novel that his profession and social position necessarily impose their own limitations of vision. In particular, he realizes that his assumption that knowledge of nature and its value can be separated from political determinations results in a kind of socioecological unconscious. In this sense, in *Get a Life* Gordimer remains as relentless in her focus on the ways that the "natural" cannot be separated from the political as she did in *The Conservationist*.

A (frightening) disruption of Paul's "ordinary" life triggers access to aspects of his environment that have been hidden from him. The novel begins with his return to his childhood home after surgery for thyroid cancer. Because he has undergone radiation treatment that has made him dangerous to others, he must go to live with his parents for more than two weeks in order to protect his young son. This experience overturns Paul's normal roles, practices, and relationships. An active, thirty-five-year-old professional, he is usually in the "wilderness" doing his work or at home helping to run his middle-class household. His body is normally an extension of his will, a fully known part of himself he can control. Now, relatively helpless, he is confined to his parents' house and garden. With this radical transformation of his life comes a break from usual ways for understanding environment.

Initially, the disruption in perception is most obvious in terms of Paul's body. He begins to think of it as a "territory," an aspect of environment, which determines the self in unknown (and uncontrollable) ways (14). With this change in his sense of his body comes a reconsideration of his identity, including his sexual identity. When he is given a massage by a male masseur, it is supposed to help him "to know that body again" (87). However, as the masseur works, Paul becomes aware of a possibility in his body he would never expect; he becomes erect as the man works on him: "That other self of a man, restored to him. Under the hands of a man." The result is a further sense of alienation from the self he has known and assumed: "So unquestioning about himself. This question coming now. Take what he is feeling as the last alienation of that state of existence" (88). The activation of the socioecological unconscious (Paul's awareness of those aspects of his body that have remained suppressed) disrupts a fundamental component of his identity, his notion of his sexuality, and results in an "alienation" from himself.

Soon after he experiences another kind of "alienation," this time from his professional identity. As a conservation ecologist, he has absolute confidence in his understanding of the ecosystems and ecological relationships he strives to protect, as well as their significance: "We work on background scientific research to make protest based on absolutely undeniable facts. Try for what's unchallengeable" (115). However, as he sits in the garden thinking about the threat to the Okavango delta in Botswana by the building of ten dams, he begins to question what he knows about nature and humans' relationship with it. His reflections begin with a conception of environmental organization that is typical for him and that challenges conventional forms of political mapping: "The Okavango is an inland delta in Botswana, the country of desert and swamp landlocked in the middle of the breadth of South West, South, and South East Africa. That's it on the maps; nature doesn't acknowledge frontiers. Neither can ecology." His professional vision enables him to see beyond the limits imposed by national "frontiers," to map the world using ecological connections, and, by implication, to think differently about political interest and action. As he thinks about the delta further, he reflects on how it challenges even his scientific knowledge: "he realized he knew too abstractly, himself limited by professionalism itself, too little of the grandeur and delicacy, cosmic and infinitesimal complexity of an ecosystem complete as this. . . . Where to begin understanding what we've only got a computerspeak label for, *ecosystem*" (90–91). Paul becomes aware of his own environmental unconscious, of how his profession and its jargon have limited his "understanding." He then calls his colleague in order to discuss how they can use this new insight to protect the delta; if they can emphasize the "inconceivably" maintained operation of this system, in particular "the beautifully managed balance" that keeps salt at acceptable levels, then they can convince others of the foolishness of the dams: "We're chronically short of water and it's not understood that . . . this intelligence of matter, receives, contains, processes, finally distributes the stuff God knows how far, linking up with other systems. . . . And some . . . consortium's going to drain, block and kill what's been *given*" (92–93). Despite his insight, Paul still imagines the delta within the parameters of his profession; he thinks of its ecosystem as having reached a form of perfection ("beautifully managed balance") in terms of the development of "life" that puts its value beyond question.

However, after he gets off the phone, his reflections take him further; he enters "areas of thought" that "question certainties" without which he cannot "go on pursuing what" he does, "being what" he is (93). Specifically, he entertains the possibility that the meaning he ascribes to the destruction of the delta may be circumscribed and may not be the truth he assumed: "Maybe we see the disaster and don't, can't live long enough . . . to see the survival solution. Matter with infinite innovation has found, finds, will find, to renew its principle—life: in new forms, what we think is gone *forever*. In millennia, what does it count that the white rhino becomes extinct" (93–94). This insight has profoundly disturbing implications for him. The good he believes he is fighting for cannot be defined in absolute terms; it is not given from elsewhere: "So, what is this kind of stuff, thinking. . . . Heresy, how can it come to one who when asked, And what is your line, answers, what am I, I'm a conservationist, I'm one of the new missionaries here not to save souls but to save the earth" (94). Transgressing the parameters of thought given to him by his profession (committing "heresy"), Paul has lost his certainty. If the earth and life on it are, in a sense, like ecosystems that find ways to renew themselves after individual parts are destroyed (only now, ecosystems themselves can be thought of as those parts), then conservation work is no longer about ecological protection as a necessary good, sanctioned by a higher force ("life") that is known through ecological science. Instead, it is about defining damage and threat in terms of human interests, albeit while drawing on the findings of ecology. In this sense, Paul's insight is similar to one offered by *The Conservationist*. He is not protecting nature, because it is in no need of protection.

Instead what he seeks to save is the nature constructed by ecology and conservation work and, by extension, the self entailed by his role as a "conservationist."

That Paul entertains these possibilities in a garden is fitting. In a sense, his doubts point to the possibility that the nature that he knows is necessarily gardenlike, shaped by social institutions and human desire, and that his own intervention in nature is part of that shaping. What he does as a conservationist ecologist is not necessarily a result of the will of nature but is at least partly determined by professional and other social factors. His presence and practices in "the wilderness" already make it not itself—a place of pure nature following a trajectory outside human history. If his knowledge gained there enables a movement beyond the environmental unconscious of most people, his break from his ordinary form of thought in the garden is the opportunity for his emergence from his own environmental unconsciousness, which includes a challenge to the very notion of wilderness.

Paul responds to his heretical thoughts by trying to incorporate them into his professional frame of reference, the discourse of ecology, and to limit the threat they represent: "himself in this garden is part of the complexity, the necessity. As a spider's web is the most fragile example of organisation, and the delta is the grandest. . . . all the waterways and shifting sand islands of contradictions: a condition of living" (94). One of the "contradictions" in *his* new "condition of living" includes the effort to come to terms with his doubt precisely through the vocabulary of his profession; this incorporation may transform those terms, but it also becomes a way to manage the doubt itself: "Always find the self calling on the terminology of the wilderness, so unjudgmental, to bring to circumstances the balm of calm acceptance. The inevitable grace, zest, in being a microcosm of the macrocosm's marvel. Doubt is part of it; the salt content" (94–95). Paul has been amazed at the way the Okavango manages salt, both a part of the delta and that which could destroy it if in excess. Through the rest of the narrative, the reader observes both the continued seeping of doubt into Paul's reflections on his "life's work" and his efforts to manage this doubt.

Despite their unsettling implications, Paul's doubts do lead him to consider new professional strategies, even if these new strategies lead to further conceptual complications. The episode in the garden suggests that Paul cannot just rely on science—on "facts"—but must turn to rhetoric, which focuses on making appeals through techniques based on the unstable, treacherous world of human interests, emotions, and values. After that episode, he thinks more often and more effectively about how to "sell" conservation using such techniques. For example, at one point in developing a plan to oppose the mining project, he tells his fellow ecologists, "Co-ordinate all the organisations and groups for action, jack up overseas support." This "jack up" of support even includes "pop stars who'll compose songs for us." At the same time, he remains aware that as the project has "desperately become like any other publicity campaign," it risks being compromised (146). To rely on the rhetoric of the ad agency potentially undermines the apparent solidity of the scientific grounding that he and his colleagues use for arguments against environmentally damaging development. Furthermore, it potentially legitimates the ad industry, which thrives on and is in league with the forms of development they oppose.

However, if Paul is aware of the contradictions involved in his efforts to use an approach shaped by the marketing industry, he remains blind through most of the narrative to the complications resulting from his lifelong privileged position as a middle-class white man in South Africa. In his final reflections on his professional activity, Paul begins to confront this issue. He returns to his "heresy" by focusing on the significance of human interest for the definition of an environmental problem and, therefore, for the meaning of his work. He is thinking again about the impact of the building of the dams in the Okavango. He does so in terms of a "human reality" that has to be understood in terms of the shifting grounds of need as determined by position and perspective, "however you're seen or you see yourself, the immediate, market reality—that's what counts in what you learn from the mother of your children . . . is the real world" (182). Paul's wife is a marketing executive, working in a world in which the focus is on the knowledge, manipulation, and production of fear and desire. For her, the pragmatic "real world" is opposed to the "innocent environment" of the "wilderness" in which he works (153). If the "reality" from which he has worked has been semidivine, the "eternal" reality of nature, hers is a limited human one: "People don't live eternity; they live a finite Now" (182).

This line of thought takes him to another environmental threat he and his colleagues are challenging. In a clear reference to the current situation in Pondoland (see the conclusion to the previous section), an Australian mining company is attempting to secure mining rights for the dunes in an ecologically sensitive area of the Wild Coast. The company has offered 15 percent of its profits to local people in order to secure this contract.[25] Paul recognizes that this deal and the toll road that will accompany it can be perceived as bringing financial benefits—the kind of benefits he already enjoys:

> No-one can disagree with the necessity for blacks to enter the development economy at a major level, fifteen percent is a good start? . . . There's also the concomitant reality that a toll high-way carrying the derived minerals and ilmenite . . . might bring a weekly wage to replace the sacrifice, God's gift of a few crop fields; unique endemism, and twenty-two kilometers of sand dunes which used to be fished from instead of mined. Bring hi-fi systems and cars. Yes! Easy to sneer at materialism and its Agency seductions while existence within it has the luxury of dissatisfac-tion, the wilderness to oppose it.
>   Who's to decide. (183)

In Paul's ultimate expression of doubt, he acknowledges the possibility that what he has always considered an objective reality, the wilderness and its importance, could be the product of his own positioning within "materialism and its Agency seductions," which generates (among its many luxuries) a possibility of escape into a supposedly pure nature. If this is the case, the basis for decisions ("who's to decide") regarding is-sues of development and conservation cannot transcend the influence of socioeconomic forces. As an ecologist, he may be able to make informed claims about the ecological impact of human action; yet arguments about the significance of this impact and about a corresponding course of action will be shaped by the "human reality" of unequal economic and political relationships and the perspectives they entail. One has not escaped the problem of social justice even (perhaps especially) when one has embraced the supposedly extra-human world of the wilderness.

At this point, Paul pulls back from his observations and returns to the "reality" he shares with his colleagues as they contemplate their next strategy. Interestingly, in this effort he turns to a gendered discourse of separate spheres—wilderness and garden, nature and culture; his he-retical reflections do not belong with his fellow ecologists "with whom is shared what the self pursues as reality. She. Benni [his wife], it must be allowed, is the other reality. . . . Hers, chosen, or advised by its effective-ness in the finite. . . . This kind of subject is left in the garden" (183). To continue his work as a conservation ecologist, he must work within an

objectively known natural "reality" from which will come the grounds
for conservation rather than within a limited ("finite") and constructed
"reality" determined by subject position. As a result, he attempts to separate the two realities once again and even draws on stereotypical gendered language to achieve this effect. Yet the continual return of his
doubt suggests a lasting change in his conception of knowledge, and this
change is marked even at the apparent moment of reversal ("what the
self pursues as reality").

The novel ends with Paul reaffirming that "final license of destruction must never be admitted, granted. That's the creed. Work to be done"
(187). The problem, of course, is that the forces of "destruction" mutate
in their effort to get around conservation laws and policies. In response,
Paul and his colleagues must adapt and accept that there will be no
final success: "Monday the four-wheel drive back to the wilderness . . .
according to the week's plan of research to which there is never a final
solution, ever. That's the condition on which the work goes on, will go
on" (169). The questioning of "finality" echoes a primary theme of the
novel as a whole—the ways that new realities are constantly being created from the transgression of meaning-making boundaries: "Success
sometimes may be defined as a disaster put on hold. Qualified" (99). In
the face of such conditions, the characters must "get" a life by adapting
and defining their reality. For the characters to continue to act effectively in the world, they must resist certainty and question seemingly
foundational categories. Yet to act one must also put a stop to the play
of meaning and identity; one must get "a life," even if its contingency is
recognized. Paul must hold on to his reality, which includes confidence
in the principles of ecology and the goals of ecological sustainability,
even if, to remain viable, it must be questioned and transformed. This
position is similar to Stuart Hall's when he claims that, if agency, "in
any specific instance, depends on the contingent and arbitrary stop—
the necessary and temporary break in the infinite semiosis of meaning,"
we must also remain aware of the way that "meaning continues to un-
fold . . . beyond the arbitrary closure which makes it, at any moment,
possible" (397). In *Get a Life* such contradictions remain an inescapable
part of life.

Not surprisingly, given this perspective, the novel's stance on the
study of ecology is contradictory. On the one hand, it embraces underlying principles of ecology. Gordimer stresses the importance of systems of relationships for the constitution and preservation of identity

and life, and she represents ecosystems themselves as among the most important of these systems. On the other hand, *Get a Life* emphasizes the *necessary* restrictions placed on the study of ecology by the forms of political organization and ideology that shape it. The novel does not suggest that the knowledge given by ecology is false, but rather that it must be understood as limited by ecology's political unconscious. In *Get a Life*, the more one uncovers "the environmental unconscious," the more one must confront a conundrum in which ecology's truths—its "reality"—cannot be reduced to an ideological effect but also in which those truths are never beyond the shaping influence of ideology. This unstable zone challenges efforts to establish any final separation or priority in terms of the relationships between ecology and politics, environment and ideology.

Get a Life's complex stance on ecology and conservation represents a substantial departure from *The Conservationist*. Paul's ecological knowledge and environmentalism, unlike Mehring's, are by no means entirely undercut by Gordimer's irony, and their significance is not restricted to the ways they legitimate unequal socioeconomic relationships. In this sense, the later novel offers a "more sympathetic exploration of environmental activism" than *The Conservationist* (Graham 195). Just as important, in contrast with the automatonlike Mehring, who cannot accommodate challenges to the self and world he knows, Paul's crisis leads him to question his reality and, to some degree, to internalize the insights he gains from that interrogation. At the same time, he remains limited by his position as a white, bourgeois, male subject. *Get a Life* emphasizes, more than does Gormider's earlier novel, the possibility of change for the conservationist figure, but not the achievement of enlightenment or transformation. However, in the later novel the meaning of this figure is still opened up, as is the significance of ecology and conservation. The activation of Paul's socioecological unconscious is not just a means of interrogating the values of the middle-class protagonist, not just the means to generate irony, but also the shifting ground for new possibilities.

Delving further into this aspect of *Get a Life* can be facilitated by considering changes in Gordimer's literary project between the publication of *The Conservationist* and the publication of the later novel. For much of her career, this project involved interrogating the consciousness bequeathed to whites by apartheid, which she saw as "the final avatar" and "ultimate expression" of colonialism in Africa (Gordimer,

Writing 133; Gordimer, "Living" 262). She sought to excavate the "hierarchy of perception that white institutions and living habits implant throughout daily experience in every white, from childhood" ("Living" 265). This project, she claimed, contributed to social change in South Africa: "The expression in art of *what really exists* beneath the surface is part of the transformation of a society" (*Writing* 131). Among the "white institutions and living habits" she saw as perpetuating colonial ideology and relationships were conservation and ecology. For her, they were inextricably tied to apartheid (particularly in terms of issues of land ownership), and the kind of false consciousness they generated needed to be undermined in order for social relations and their implications ("what really exists") to become apparent.

The years following the end of apartheid have transformed Gordimer's project. She continues to explore the forms of consciousness instilled by apartheid and its legacies; however, she necessarily examines them in terms of a different political landscape. If vast inequalities based on race continue, these inequalities are no longer enforced by law, and the relationship between class and race has become substantially more unstable. Furthermore, if the state continues to protect entrenched economic interests, it cannot ignore, at least in the same way the apartheid government did, the interests of the impoverished majorities. Finally, as South Africa has become like other liberal, democratic nations, the ways socioeconomic identities within the nation are linked with global capitalism have been highlighted. In *Get a Life,* Gordimer is more interested than she was in her apartheid writing in class identities that span the globe and in the power of global capital to shape the nation. In her depictions of foreign companies working with local entities to wrest control over development, Gordimer seems to be in some doubt that she saw the end of imperialism with the death of the apartheid state. In this situation, if the perspectives of privileged classes often coincide more with those of elites in other nations than with those of the majorities within their own nation, concern with threats to self-determination, economic well-being, and the right to a healthy environment posed by the operation of global capital can still transcend lines drawn by class, race, and ethnicity.

The shift in Gordimer's project resulting from her attempts to address changing political circumstances impacts her conceptualizing of conservation and ecology in *Get a Life.* She takes into account that the natural environment has become "a concern across social sectors" and, as a result, is "no longer viewed as the preoccupation of an elite rooted in a colonial past (Vital, "'Another'" 97). She gestures toward the ways environmental work and legislation can be influenced by the needs of a greater percentage of the population and can combat development projects that exploit both people and places: "The liberalization of South African politics created discursive and institutional space for rethinking environmental issues, and a vibrant debate on the meaning, causes, and effects of environmental decay began in earnest" (McDonald, "Environmental" 257).

At the same time, Gordimer remains alert to the ways environmentalism "might relate in discomforting ways to economy" (Vital, "Another" 98). She represents it as still shaped by an environmental discourse that enables conservationists from the privileged classes to align themselves with nature while ignoring and/or perpetuating the forms of economic development that have given them their advantages and that are often the underlying causes of the environmental threats against which they struggle. As David McDonald notes, "mainstream environmental groups in South Africa have been blamed for not taking environmental degradation in the townships and former homelands seriously" and for a paternalistic emphasis on "the need to 'educate' black South Africans about the environment, or the need to protect wildlife reserves from the 'population explosion' taking place in the rural (read black) areas surrounding national parks" ("Environmental" 261–62). Meanwhile, white environmentalists are part of a population contributing disproportionally to many of the major environmental threats of our time: "Middle and upper-income South African households are among the most wasteful users of resources in the world. Per capita consumption of water, electricity, and other basic resources in historically white suburbs . . . is as high or higher than in any country in the world" (275).

Gordimer represents her ambivalent take on conservation in South Africa using Paul's character. While she depicts him as sympathetic and his work as valuable, he also points to the continued limitations imposed on ecology and environmentalism by bourgeois subjectivity. Anthony Vital argues that, despite her ambivalence, Gordimer ultimately uses Paul to suggest that, against the backdrop of global capital, his form of conservation "may indeed appear as not much more than a form of middle-class coping" ("Another" 107). In this reading, Paul remains a type of privileged ecologist and conservationist unable to think outside an ideology of separate spheres (nature/society, private/public).

His "suburban form of ecological thinking," Vital claims, is revealed to be "profoundly ideological, a thinking in the (unacknowledged) service of maintaining the social order that benefits him" (102). His complicity with the forms of predatory capital prevents him from contributing to any lasting change. Gordimer, according to Vital, depicts Paul's form of conservation as lacking any "particular socially transformative vision," and as a result, it takes "on for the comfortable classes the character of a seemingly endless defensive action" (106).

While this reading of the novel's perspective on Paul and his work is powerful, it does tend to present him as static (as a type). Focusing on the theme of the socioecological unconscious in *Get a Life* and on Paul's incrementally changing perspective yields a somewhat different, more optimistic reading of his character and its significance. If, as a result of his commitments, the changes in his life, and the subsequent activation of his socioecological unconscious, he cannot permanently suppress a growing awareness of the impossibility of separating conservation and social interests, as well as ecology and rhetoric, then his work and even some of the institutions with which he is associated have the potential to be transformed and could become even more useful in creating a better future. *He* is part of the current circumstances from which a new order might be born, unlike Mehring, whose *absence* at the end of *The Conservationist* is one means by which the text gestures toward a better future arising from the circumstances of the present.

4 THE NATURE OF Violence

KEN SARO-WIWA FAMOUSLY CHARACTERIZED GAS FLAR-
ing and oil spills in the Niger Delta as a form of genocidal violence.
His manifesto *Genocide in Nigeria* (1992) claimed that the Ogoni people
were left "half-deaf and prone to respiratory diseases" and that their
main livelihoods, farming and fishing, were being destroyed by the poi-
soning of air, water, and soil (81–82). Meanwhile, they saw almost no
benefits from the oil pumped from their land and lacked basic infrastructure like electricity, health care facilities, and schools. This situation, Saro-Wiwa argued, was caused by the willful negligence of the international oil industry and the Nigerian government and evolved from
Nigeria's development along (neo)colonial lines.

Through his leadership in the Movement for the Survival of the
Ogoni People (MOSOP), Saro-Wiwa came to pose a significant threat
to the Nigerian petro-state and the economic interests it protected. Michael Watts characterizes the petro-state in terms of a contradiction between claims to legitimacy based on the power of oil profits to develop
the nation and the actual deterioration engendered by the "slick alli-
ances" of the state and the oil industry. If oil strengthens "the centrality
of the nation-state as a vehicle for modernity, progress, civilization," it
also "produces conditions that directly challenge and question those . . .
tenets of nationalism and development" and that reveal "the state and
the nation to be sham, decrepit, venal, and corrupt notions" ("Petro-
Violence" 208). As described by Andrew Apter, the historical trajectory
of the Nigerian petro-state is a particularly salient example of this process. The oil boom of the 1970s created the appearance of development,
but the nation's means of production atrophied: "The oil economy . . .
intensified [the] circulation of money and commodities, but it enervated
and undermined the real productive base of Nigeria, those agricultural
resources that not even a state-sponsored green revolution could revive"
(269). At the same time, the state increasingly became a means for the
accumulation of private wealth based on "a pattern of patronage in
business and politics that allocated licenses and revenues in exchange
for kickbacks and loyalty" (264). As the oil boom of the 1970s withered

away and Nigeria's economy suffered under the fluctuations of international commodity markets, the contradictions of the petro-state became more apparent; huge private fortunes continued to be made by a small, corrupt elite as civil society deteriorated. An intensified use of violence to keep popular protest in check made the divorce between the state and the people especially stark: "There was no sphere of *res publica* in Abacha's Nigeria; no effective system of interest articulation, legal process, public education, press coverage, or publicity, nor was the most basic protection of life and liberty even recognized by the state" (272). In other words, Nigerians were not citizens but (neo)colonial subjects. At the same time, oil extraction ruined the land and waterways and, in the process, polluted and destroyed "the productive base of the ecosystem" (273).

Apter argues that MOSOP's campaign had particular resonance in Nigeria by the time of Saro-Wiwa's death because it was trying to reclaim and save "the very ground of civil society itself": "as citizens took to the streets in defense of their citizenship, the Ogoni struggle joined hands with a larger national cause. Thus the ecological destruction . . . in the remote areas of the Niger Delta epitomized the pollution of the public sphere by an invasive and extractive petro-state" (275–76). With environmental degradation in the delta serving as a potent symbol of a Nigerian state destroying the foundations of the nation, the people *and* the land, the Ogoni's struggle for environmental justice turned into an iconic example of popular dissent against the criminal and illegitimate state.

After Saro-Wiwa's execution by the state in 1995, his transformation into a potent symbol for resistance against petro-capitalism became even more pronounced. His ability to work within and across different scales has been especially influential. Yet escalating violence in the Niger Delta over the last two decades—slow and fast, ecological and social, disorganized and organized—points to a need not only to reiterate but also to revise Saro-Wiwa's resistance narratives. Critical commentary generated over the past twenty years does not necessarily suggest that those narratives and his inspiring example are of no further importance or use, but it does foreground the importance of thinking critically about how they are deployed and how they can be made relevant to the present. As Rob Nixon suggests, the kind of mythic figure Saro-Wiwa has become "can become a powerful political asset but also stands dauntingly in the path of those who wish to take the struggle forward in new ways, for new times" (*Slow* 123). Graham Huggan and Helen Tiffin also worry about Saro-Wiwa's "metonymic function as a global spokesperson for social and environmental issues" that "can easily lead to moralistic generalisations about endemic political corruption in Africa . . . or the heroic part played by freedom fighters and resistance movements prepared to take on the assembled might of global commerce and the state" (41). In other words, we need to move beyond celebrating Saro-Wiwa as an icon of African environmental imagination and upholding his stories as a framework for environmental justice struggle locally and globally.

Reading *Genocide in Nigeria* and *A Month and a Day: A Detention Diary* (1995) as part of an anticolonial environmental narrative tradition and considering how other authors deal with imperial development along the lower Niger are useful means to reinvigorate discussion about Saro-Wiwa's writing and the issues he raised regarding socioecological justice in the Niger Delta. For example, while Saro-Wiwa's narratives about development in Nigeria have some striking parallels with Chinua Achebe's *Arrow of God,* exploring differences between Achebe's and Saro-Wiwa's representations of place and place-based identity can enrich debate about such representations and their significance. In a more contemporary context, the poets Tanure Ojaide (*Delta Blues and Home Songs* and *The Tale of the Harmattan*) and Ogaga Ifowodo (*The Oil Lamp*) draw inspiration from Saro-Wiwa's counterhegemonic narratives; however, in their departures from his assumptions regarding ethnic identity and the grounding for resistance, they enable a reconsideration of those narratives in relation to current conditions.

Writing Resistance: Ken Saro-Wiwa

Regularly harassed and detained by the Nigerian petro-state, Saro-Wiwa was eventually arrested for the murder of four Ogoni chiefs by a mob during a MOSOP rally. Even though he was far from the scene of the crime, and despite the fact that he unremittingly advocated the use of nonviolence in the Ogoni struggle, he was charged with inciting the crowd through his speeches and activism. Orchestrated by Nigeria's new military dictator, Sani Abacha, this detainment was to have a different outcome from the previous ones. After being held for months without charge or trial, Saro-Wiwa was brought before a military tribunal handpicked by Abacha. The subsequent trial was a travesty. Prosecution witnesses admitted being bribed; the defense team was regularly harassed and threatened by security agents; and eventually "the bias of the tribunal was so blatant that the defense team withdrew, declaring that their continued participation would only give a semblance of legality to a patent circus spectacle" (Soyinka, *Open* 146). On October 31, 1995, Saro-Wiwa was sentenced. In the London *Guardian,* Wole Soyinka noted that with the subversion of the judiciary and the creation of secret tribunals by Sani Abacha and his henchmen, the verdict "was, of course, only too predictable. Abacha had decreed the death sentence for Ken Saro-Wiwa, and nothing else" (*Open* 147). Despite an international outcry, Saro-Wiwa and the eight other MOSOP activists arrested with him were hung to death on November 10. In the process, the Nigerian state showed itself willing to use any available means to silence its critics and instill fear by demonstrating its unrestrained "power of life and death over its citizens" (Quayson 72).

Yet Saro-Wiwa's death by no means put an end to his influence. In fact, Abacha had created a martyr "invested with a mythic quality" whose story was turned into "a morality tale for the late twentieth century" (Wiwa 2). There was a huge international outcry: "in death, Saro-Wiwa extended . . . the remarkable coalition of international interests that he had begun to forge while alive, an alliance that brought together environmentalists, minority rights advocates, anti-racists, opponents of corporate deregulation, and defenders of free speech" (Nixon, *Slow* 122). As Saro-Wiwa anticipated, his memory also continued to inspire the Ogoni. In a letter to his son, he proclaimed that "in the event of death" he would "pass into Ogoni folklore" (Wiwa 127). This prophecy proved accurate; Ken Saro-Wiwa's story "was a crucial chapter of Ogoni history . . . that was supposed to be told and retold, embellished and mystified with every retelling" (Wiwa 11). More generally, proclaims Wiwa, his "father's life and death" inspired "millions of people around the world who are struggling for social justice and human rights" (11). Finally, Saro-Wiwa has remained a rallying point for those trying to make Western environmentalists more attuned to the concerns of peoples they have often ignored and especially to ecological catastrophes in Africa having devastating implications for humans (but not involving charismatic megafauna). As Rob Nixon notes, "Saro-Wiwa's campaign for environmental self-determination may well prove historically critical to the development of a broader image of ecological activism" beyond not only Europe and North America but also "Amazonian and Indian examples" (*Slow* 112).

Despite Saro-Wiwa's continued influence, conditions in the Niger Delta have only grown worse since his death. A UN report published in July of 2011 documented no change in terms of environmental devastation, its social effects, or the oil industry's efforts to address them (Dixon). The Niger Delta remains one of the most polluted places on earth and would take thirty years and one billion dollars to clean up (if the industry could be held accountable). As the report notes, the oil companies still fail to follow even the most basic procedures for maintaining their operational infrastructure and for environmental protection. Meanwhile, they stick to their claims that they operate according to the highest standards of business ethics and that oil spills are the result of criminality among the peoples of the Niger Delta.

There are numerous reasons for this lack of substantive change in industry practice and rhetoric. First, the companies and government remain extremely dependent on profits from extraction in the delta. Nigerian oil and gas remain "core assets in Shell, Chevron and Exxon-Mobil's global portfolios," and they continue to account for more than 80 percent of the Nigerian petro-state's revenues and 40 percent of Nigeria's GDP (Rowell, Marriott, and Stockman 96–97). Given these circumstances, it is little surprise that the relationship between the national government and the multinational oil corporations remains as cozy as ever. In addition, in the wake of 9/11, Nigeria became central in the Bush Administration's plans to decrease the United States' reliance on oil from the Persian Gulf.[1] The concern with "energy security" put a premium on protecting and increasing production in the delta, while further deemphasizing the importance of human rights, environmental justice, or "a de-militarized sustainable post-oil future" (Rowell, Marriott, and Stockman 216). In turn, the American military's relationship with their Nigerian counterparts grew closer, as did cooperation between the US government and the oil business in Africa. Finally, the industry, with the help of a compliant media, has been able to mount extremely effective PR (Rowell, Marriott, and Stockman 115–29).[2]

The two decades following Saro-Wiwa's death have also included a "calamitous descent into violence" in the Niger Delta (Watts, "Sweet" 38). Continued ecological and social devastation and vicious state-sponsored terror fuel both "a gigantic reservoir of anger and dissent"

(39) and a sense of frustrated hopelessness. Armed rebel groups have proliferated and turned to kidnapping oil workers and sabotaging installations. From late 2005 until 2007, a union of these groups, the Movement for the Emancipation of the Niger Delta (MEND), compromised 40 percent of Nigeria's oil revenues. Creating even more havoc, armed gangs and ethnic militias fight for control of the oil-bunkering trade or for protection money from oil companies. This money is tied to the companies' efforts to channel conflict for their own purposes and instrumentalize inter- and intracommunity groups.[3] Meanwhile, the Nigerian military, even after the return to democratic rule in 2000, has continued to work closely with the oil industry. What Claude Ake called the "militarization of commerce" (qtd. in Nixon, *Slow* 107) and the "privatization of the state" (qtd. in Okonta and Douglas 60) remain significant as the companies get protection both through the intervention of the government and through direct payments and logistical support for military working in the delta (Rowell, Marriott, and Stockman 15, 99).

These recent developments point to some limitations in Saro-Wiwa's resistance narratives. Particularly problematic is his claim that MOSOP was grounded in a historically transcendent Ogoni "genius" enabling the shedding of false consciousness and of differences among the Ogoni generated by (neo)colonialism and his concomitant suggestion that, if allowed, MOSOP would necessarily produce a truly decolonized semi-autonomous state functioning in the interests of all Ogoni and freed from the injustice and violence plaguing Nigeria (*Month* 130).[4] This notion potentially suppresses the degree to which identity was shaped throughout Nigeria by colonial modernity and by the development of the petro-state. It also sidesteps the extremely tricky question of whether and how compensation in the form of petrodollars can avoid becoming a corrupting rather than a constructive force given infrastructural, political, and discursive conditions in the delta and Nigeria as a whole.

In *A Month and a Day* and *Genocide in Nigeria*, Saro-Wiwa effectively represents the historical causes working at interrelated geographical and temporal scales for devastating socioecological transformation in the Niger Delta. His focus on the politics of scale, his challenge to official development rhetoric, and his counternarrative representing oil extraction as annihilation foreground the relationship between these texts and *global* environmental justice discourse.

In Saro-Wiwa's counternarrative, Nigeria's imperial trajectory has discouraged unity, encouraged conflict among groups, fostered destructive attitudes toward resource allocation and the natural world, and suppressed the possibility for alternative models of development. *Genocide in Nigeria* begins with a snapshot of a unified, autonomous, sustainable precolonial society; the Ogoni "were able to preserve their land and culture and maintain peace and order in their territory in virtual isolation" (14). He anchors their Edenic way of life in the Ogoni's ecologically sensitive animism. Their worship of "the land" resulted in forests being viewed as "not merely a collection of trees and the abode of animals but also, and more intrinsically, a sacred possession" and in "rivers and streams" being more than just resources, "water for life—for bathing and drinking," but also "sacred" and "bound up intricately with the life of the community, of the entire Ogoni nation" (12–13). According to Saro-Wiwa, the Ogoni's spirit of self-determination, cultural unity, and connection with their natural environment led them to be marginalized in an imperial dispensation. Refusing to cede "their sovereignty," they were considered backward (15); "the area was left to stagnate," other ethnic groups benefited at their expense, and their interests were not considered "in the burgeoning politics" of colonial Nigeria (15, 19). Embracing values encouraging protection of the land and resistance to external control, they and their way of life were devalued and set up to be sacrificed.

Saro-Wiwa also stresses how Nigeria's "regional arrangement" based on "the preponderance of one ethnic group" in each region led not only to conflict among the three major groups but also to the Yoruba, Igbo, and Hausa-Fulani becoming "the power brokers in Nigeria, with the minority ethnic groups in each Region attached to them as mere appendages" (20). This situation hampered the growth of any sense of Nigerian identity and common national interest; primarily loyal to their ethnic groups, "most Nigerians pay only lip service to the concept of the Nigerian nation state" (84). Since independence, this arrangement has led to struggle among the major ethnic groups for the loot of the nation and internal colonialism for small groups, in which their interests are ignored and they are positioned as disposable in the name of progress. In other words, the language of national or regional development functions transparently as neocolonial discourse in Nigeria, legitimating exploitation through a baneful movement from metonymy to metaphor in which the progress of some can stand in for the progress of all.

Because of the huge reserves of oil lying beneath their land, the

Ogoni were hit especially hard. Politically marginalized at regional and national levels, they received little compensation for oil exploration and extraction. The government siphoned off the wealth extracted from Ogoniland to create private fortunes for the ruling elites and to develop other parts of Nigeria, even as the Ogoni themselves lack all basic infrastructure and amenities such as electricity, potable water, roads, schools, and hospitals. As Saro-Wiwa emphasized in *A Month and a Day,* the situation became especially severe after 1977. A new Nigerian constitution was imposed by the military junta, which declared all land in Nigeria the property of the federal government, strengthened the central government, and "left the ethnic minorities totally unprotected in terms of their economic resources and their culture" (42). The result was that "by 1980 the Federal government had left the oil-bearing areas with only 1.5 per cent of the proceeds of oil production" (42–43).

Echoing Frantz Fanon's analysis of the "pitfalls of national consciousness," Saro-Wiwa also explains how the relationship between the Nigerian state and the oil industry is the continuation of a dynamic begun under colonialism. Created as a tool to further "British economic interests," the colonial constitution gave "the Crown" control over "mineral rights and publicly purchased land" (*Genocide* 21). In turn, the colonial administration issued "an oil mining lease" to Shell-BP, enabling the company to take land on which oil was found for little or no compensation and, more generally, "to do whatever it pleased in the search for oil" (24). This use of political power undermined "customary law," which "prevented economic exploitation of the land by individual Nigerians" (21). After independence, the oil companies continued to appropriate land and avoid regulation by working closely with the Nigerian national government. Basically, they maximized profits at the expense of Nigeria as a whole, and especially its ethnic minorities, by enriching a small ruling elite who served as middlemen and enforced the interests of foreign capital. In Fanon's terms, these elites took "on the role of manager for Western enterprise" even as they espoused nationalist rhetoric (154). In *A Month and a Day,* Saro-Wiwa uses a vivid metaphor of a masquerade to describe this process; with "black colonialists wearing the mask of Nigerianism," there was a new masquerade "leashed to a rope held by an unseen hand, and steadied by the oil of the Ogoni and other peoples in the Niger River Delta. In effect, the producers of that oil, the multinational oil giants, truly controlled the masquerade in the arena" (126). This "masquerade" resulted in a lack of even the most basic regulations for the oil industry in Nigeria and, ultimately, in an imperial geography of environmental injustice. The companies follow strict guidelines in the United States and Europe while the ecological homes, health, and livelihoods of communities in Africa are destroyed without compunction: "Shell has won prizes for environmental protection in Europe where it also prospects for oil. So it cannot be that it does not know what to do" (*Genocide* 82).

A significant portion of *Genocide in Nigeria* focuses on challenging what Saro-Wiwa calls "shellspeak," that is, the kind of (neo)colonial development discourse that has "imperialized the wishes and world-views" of the wealthy and powerful (Escobar 194). Shell had historically claimed both to bring economic progress to the delta and to follow the strictest environmental guidelines in its operations (Okonta and Douglas 63). Foregrounding local conditions that have been rendered invisible and drawing on the voices of other Ogoni activists from the past, Saro-Wiwa undermines the company's rhetoric of care and its efforts to obscure and sooth.

Responding to Ogoni complaints in 1970, Shell claimed both that it was making a "significant contribution to the overall economic development" of Nigeria and that it was "extremely careful to ensure that . . . operations cause minimal disturbance to the people in the areas in which [they] operate" (Genocide 51). Soon after this public letter was sent, "there was a major blow-out at the Bomu oilfield," caused by Shell's "negligence in installing outdated equipment" (57, 80). In protest, various youth groups wrote letters to Shell cataloging the environmental devastation from the blowout and, more generally, from oil exploration and extraction. They claimed that rather than bringing development, Shell had destroyed their communities and their futures. While the company "blows such a loud trumpet about the assistance she renders to peoples in her areas of operation," and while the letter writers at one time believed that Shell would bring "prosperity, good health, good education, employment opportunities," they "know now that [they] were celebrating [their] entry into a darkened and oblique horizon of despondency, abject poverty, extinction of . . . lives and destruction of . . . crops" (65). Saro-Wiwa also includes a letter from a group of "Ogoni Citizens" to the publishers of the "weekly *West Africa*" claiming that inattention to the blowout is part of a larger process in which the material misery produced by mineral extraction is masked by the numerical abstractions of economic growth, the "worthless and endless computations of how many millions Nigeria is due to get in foreign exchange earnings from petroleum" (71). Finally, Saro-Wiwa documents Shell's efforts to recover its lost oil and its relative lack of concern for "the sufferings of the people whose food crops, economic trees and farmlands [had] been destroyed and fishing waters polluted by the blow-out" (72).

The conclusion of this section, a catalog of damage done to the Niger Delta's ecosystem and to the health and livelihoods of the Ogoni by the flaring of gas and oil spills, serves as a means for Saro-Wiwa to launch his central claim: Shell's and Chevron's actions "amount to genocide. The soul of the Ogoni people is dying and I am witness to the fact" (83). This accusation is grounded in the documenting of the Ogoni's reliance on the delta's ecosystem, of unfolding socioecological disaster, and of that disaster's connections with (neo)colonial development and development discourse. It entails a shift in the conception of intentional, targeted violence against an entire collective to include the effects of toxins generated by willful neglect. In the context of Saro-Wiwa's counterhistory, the potentially hyperbolic term *genocide* becomes a strategic means to foreground the scope of disaster and injustice he has witnessed, and it serves to counter the much more implausible rhetoric of development deployed by the oil companies and the Nigerian state. It helps position such development and the interests it serves as unimaginable horror and, in this sense, is a crucial component of a postcolonial environmentalist rhetoric of crisis and its causes.

While reiterating the history lesson of *Genocide in Nigeria,* Saro-Wiwa's prison diary *A Month and a Day* is primarily focused on the growth of organized resistance among the Ogoni and on what needs to be done to foster the struggle at and across different geographic scales. According to Saro-Wiwa, the development of resistant identity required first and foremost a break from the forms of consciousness instilled by (neo)colonial history and discourse. He represents MOSOP's evolution as a model of communal development based on the reclamation of a distinct transhistorical Ogoni consciousness that liberated them from a historically imposed sense of inferiority and helplessness. Their oppression has included not only "an outrageous denial of rights" and "a usurpation of [their] economic resources" but also an educational system that denigrates their culture and characterizes "them as meek, obscure and foolish" (50). The resulting loss of collective confidence led, Saro-Wiwa claims, to an acceptance of their own colonized condition:

"The Ogoni people have virtually lost pride in themselves and their ability, have voted for a multiplicity of parties in elections, have regarded themselves as perpetual clients of other ethnic groups and have come to think that there is nowhere else to go but down" (50). Faced with such conditions, Saro-Wiwa called on the Ogoni to establish a sense of "unity, unanimity, and consensus" (49) and to embrace a transhistorical Ogoni identity and culture. He claims that history reveals them to have "always been fierce and independent" (49). If they rely on their "age-old traditions" and their "genius," they can, he proclaims, "extricate [themselves] from the quagmire in which [their] abundant wealth has paradoxically placed" them (149). Saro-Wiwa thus embraces the pastoral notion of a better future to be found in a return to an identity from which the Ogoni have been violently wrenched. According to Saro-Wiwa, MOSOP helped the Ogoni access this identity and fostered the values of cooperation and communal interest through strategies such as "The One Naira Ogoni Survival Fund." Through the token amount of one naira, the fund emphasized "not money but the symbols of togetherness, of comradeship, of unity of endeavor"; contribution to the fund was "a statement of [the Ogoni peoples'] will to survive as individuals and as one indivisible nation" (99).

In many ways, Saro-Wiwa's model of mobilization was profoundly vertical. He claims that only after the Ogoni elite had "taken the first important step in clearing [their] minds" could the people be awakened "from the sleep of the century" induced by their "debilitating poverty" and their psychological colonization (75, 16, 149). Yet his narrative also has a strong egalitarian element. He suggests that leaders of anticolonial resistance must recognize that power and progress come from the people and that they must be heard (as well as instructed). Putting his literary and media skills to work for the people required direct engagement with them. In Africa, he proclaims, "the writer must be . . . the intellectual man of action. He must take part in mass organisations. He must establish direct contact with the people and resort to the strength of African literature—oratory in the tongue. For the word is power and more powerful is it when expressed in common currency" (55). Such engagement, Saro-Wiwa suggests, transformed him. He became aware of a "solidarity," "courage," and "level of expertise" among "the young Ogoni" that he had not "credited them" with before (16, 77). More generally, the determination among the Ogoni he witnessed at the huge protest in Bori on January 4, 1993, changed his perception of them and strengthened his own resolve: "I saw a new profile of the Ogoni people, a profile I had not identified. . . . Ogoni would surely not be the same again. And I also felt that I must not let them down ever" (88). Thus, Saro-Wiwa embraces the role of the Fanonian intellectual who, joining "[the people] in that fluctuating movement which they are just giving shape to," becomes "an awakener" but is also shaken and led by the people (227, 223).

A Month and a Day is particularly concerned with how the past and present point to what will need to be done in the future to strengthen MOSOP. Saro-Wiwa emphasizes how his experience with the oppressive institutions of the Nigerian state can serve as an important lesson. The Ogoni must, he claims, prepare themselves for the means that will be used to silence them:

> We had read of the detention of people in Nigeria, but it was mostly a phenomenon of Lagos, where there were several human rights activists. That was until I told the Ogoni people that they were being cheated, denied of their rights to a healthy environment, and the resources of their land. Then almost the entire 51,000 Ogoni men, women and children became activists. Still, prison seemed far away. . . . And yet I do remember that I kept warning the Ogoni people to prepare for harassment, imprisonment and death.
>
> Altogether, it was fitting that I should be one of the first to be detained. It would show subsequent detainees that they were in good company. (158–59)

The writing of *A Month and a Day* serves as part of Saro-Wiwa's lesson; it is a means to show the people that they can successfully resist, even in prison. As Ngũgĩ wa Thiong'o points out, when the artist successfully helps create a liberatory narrative, especially when this narrative is being made with and among the people, the state both performs its power and attempts to control the performance of the artist through detention, exile, and/or death. Yet, as Ngũgĩ also points out, even in prison, the narrative battle between state and artist is not over: "Prison narratives by artist-prisoners are essentially a documentation of the battle of texts and of the continuing contest over the performance space of the state. This contest, while aimed at the groups of interested watchers outside the gates—Amnesty International, International Pen, and other human rights groups—is ultimately aimed at the real audience: the people waiting in the territorial space" (*Penpoints* 57). The epitome of this type of prison narrative, *A Month and a Day* refuses the state's efforts to control representation and denies it "a triumphant epilogue to its performance" (Ngũgĩ, *Penpoints* 57).

Saro-Wiwa's prison diary is clearly "aimed" not only at the Ogoni people but also "at the group of interested watchers outside the gates" including human rights organizations like Amnesty International, also environmental groups like Greenpeace, other oppressed minorities, and political activists across Africa. He assiduously worked to find common ground between MOSOP and other political movements in order to garner support and create solidarity. As Rob Nixon notes, "Saro-Wiwa appreciated the improbability of converting an injustice against a small African people into an international cause. His strategic response was to scour the wider political milieu for possible points of connection" ("Pipe" 113). For example, in positioning MOSOP as "a model for other small, deprived, dispossessed and disappearing peoples," Saro-Wiwa appeals both to activists fighting for marginalized minorities around the world and to Western groups focused on human and democratic rights: "A large number of communities ready to take their fate into their hands and practice self-reliance, demanding their rights nonviolently, would conduce to democracy and more politically developed peoples" (92). He also appeals to environmentalists through the trope of the ecological heritage site; his "overall concern is for the fragile ecosystem of the Niger Delta—one of the richest areas on earth," and he considers "Shell's despoliation of the area as a crime to all humanity" (112). Through such adept rhetorical moves, Saro-Wiwa evokes a language of shared interests without suppressing the significance of social and cultural difference.

The success of such strategies was reflected in his ability to enlist the support of a wide variety of groups, many of which did not initially see his concern with the Ogoni's plight as linked with their agendas. For example, when he contacted Greenpeace and Amnesty International, the former were uninterested in his cause because they did not "work in Africa," and the latter were only interested in "conventional killings" as opposed to deaths caused by environmental degradation (61). Eventually, he won them over and, in the process, contributed to their transformation. As Nixon points out, by encouraging a decentering and diversifying of environmental agendas, Saro-Wiwa has helped move the environmental movement away from too strict a focus on (white, Western) concerns such as "purity preservation and Jeffersonian-style agrarianism" ("Environmentalism" 243).

Of course, the process of transformation worked both ways. If he helped change the agendas of Western environmental groups, they also helped him to reenvision MOSOP's work. He claims that a "visit to the United States sharpened [his] awareness of the need to organise the Ogoni people to struggle for their environment. One visit to a group in Denver, Colorado . . . showed what could be done by an environmental group to press demands on government and companies" (54).

Yet some care needs to be taken regarding the significance of this anecdote for understanding the origins of MOSOP's environmentalism. Saro-Wiwa's primary emphasis is on how the trip to the United States sensitized him to the strategic benefits of positioning MOSOP as an environmental movement. He later claims that "what the trip did was to convince [him] that the environment would have to be a strong *plank* on which to base the burgeoning" MOSOP (54; emphasis mine). In other words, the experience led to a recognition that *framing* the movement as environmental would help strengthen MOSOP internally and would galvanize support externally. This strategic emphasis does not negate the significance of what Saro-Wiwa learned from his cosmopolitan experiences, but it can serve as a reminder that Ogoni resistance already had a strong environmental component. In this same section describing his trip to Colorado, Saro-Wiwa stresses that his own environmentalism stretches back at least to the Biafran war and to "the blow-out on Shell's Bomu Oilwell 11 in 1971" (54). The reference to Bomu also points to the many letters discussed earlier in which various Ogoni groups focused explicitly on the ecological disasters wrought not just by the blowout but also more generally by Shell-BP's operations among the Ogoni. In other words, the "plank" of environmental violence was already a strong component of Ogoni grievance and resistance. It was shaped by the conditions created by oil extraction in the delta and a particular history of internal and international imperial relationships. What made MOSOP's environmentalism different from the kind Saro-Wiwa observed among the "group in Denver, Colorado interested in the trees in the wilderness" is precisely its foregrounding of the impact of such relationships on both people and environment. There is not a focus on a landscape supposedly unshaped by human intervention ("the wilderness") but on a populated place where livelihoods and health have been shaped by a long history of mutual ecological and social interpolation.

Saro-Wiwa claims that ultimately MOSOP's environmentalism is rooted in the ecologically oriented spirit of the Ogoni people: "A bit of research and thinking of my childhood days showed me how conscious of their environment the Ogoni have always been and how far they went in an effort to protect it. I had always felt part of that consciousness myself" (54). This comment, of course, echoes the representation of the Ogoni as eco-Indigenes offered at the beginning of *Genocide in Nigeria*. Ken Wiwa has argued that this aspect of his father's writing and activism must be thought about in terms of its strategic value rather than dismissed using the standard of some ideal, objective historical account. He claims that when his father "insisted that the Ogoni had lived in harmony with their neighbors and the environment until the Europeans arrived on the scene, he knew it was myth. It was a romantic notion . . . that was supposed to fire the individual imagination and the collective quest for cultural identity and survival." In other words, Saro-Wiwa was focused on the formulation of a usable past, that is, on "the powerful act of appropriating the past through imaginative understanding" for the sake of "the sanity of the whole community" rather than "through a 'scientific objectivity' which tries to mask its own uncertainties" (68–69). In these terms, the effectiveness of Saro-Wiwa's "romantic" narrative must be understood in relation to his successes at instilling cultural pride, undermining notions of progress associated with colonial modernity, facilitating identification with a local nature in the process of being destroyed, and encouraging alternative models of socioecological practice.

Yet, even if one measures the value of Saro-Wiwa's deployment of the eco-Indigene narrative not in terms of its truth but in terms of its strategic or pragmatic implications, it has some significant limitations. For example, it too easily elides the question of what needs to be done in the present to overcome environmental attitudes and practices shaped by colonial modernity. Saro-Wiwa implies that once members of MOSOP tap into their Ogoni identity, they will automatically be on the side of nature and against the kind of imperial practices associated with its abuse. This suggestion curtails a close examination of the specific kinds of socioecological perspectives and projects needing to be fostered and those needing to be questioned by MOSOP members (even after they have identified with the struggle for Ogoni political rights).

In turn, it occludes the troubling issue of what will happen to the environment if a semiautonomous Ogoni state is created. MOSOP's and Saro-Wiwa's ultimate goal was increased self-determination within a Nigerian confederation. The Ogoni Bill of Rights, presented to the Nigerian government in 1990, called for the Ogoni to "be granted Political Autonomy to participate in the affairs of the Republic as a distinct and separate unit" (*Month* 48). Saro-Wiwa believed that such autonomy was a means to bring about improved civil liberties and rights, economic and cultural development, better environmental protection, and a more democratic Nigeria. Yet such a state would require the reconciling of these not entirely compatible aims. In particular, "the right to control the use of a fair proportion of Ogoni economic resources for Ogoni development," in part by entering into contracts with oil companies, obviously could undermine the protection of the delta's ecosystem (*Month* 48). Reconciling the two goals clearly would entail determining what sustainable progress might mean. Even if the Ogoni state was democratic and uncorrupt (which assumes the transcendence of material and ideological conditions shaped by Nigerian history), determining a definition of environmental sustainability through communal debate and discussion would by no means necessarily result in a decrease in environmental degradation necessary to save the Ogoni environment. In a hypothetical scenario, a majority of voices could be more focused on concerns with compensation and the development of Ogoni infrastructure and amenities and could take a backseat to environmental restoration and protection. Saro-Wiwa's representations of Ogoni culture would suggest that this hypothetical scenario ignores the ways that institutions and regulations shaped in an Ogoni state would necessarily ensure such restoration and protection. However, the problem is precisely that Saro-Wiwa—relying on a nationalist notion of collective "genius" or "spirit" (*Month* 130)—never seriously considers that culture and identity might have been fundamentally transformed in such a way as to necessitate a reformulation in the present.

In general, Saro-Wiwa's naturalizing representations of transhistorical ethnic identity and indigenous belonging elide difficult issues regarding the impact of Nigeria's historical development on consciousness and material practice. He portrays a *singular* Ogoni identity from the past, embodying a collective (national) essence, which has been revived and inhabited in the present by MOSOP; this identity is utterly different and separated from the forms of consciousness and the values generated by colonial modernity. In this sense, his representations project a relatively straightforward dichotomy between, on the one hand, the oil companies and the Nigerian state and, on the other hand, MOSOP. Saro-Wiwa also suggests that a semiautonomous Ogoni state would, by virtue of its origins in MOSOP (and so in the spirit of the Ogoni people), avoid the corruption, oppressive relationships, divisions, and violence he associates with the Nigerian state and the colonial setup of Nigeria.

Bringing into question this aspect of Saro-Wiwa's narrative of resistance, Ato Quayson has suggested that the killing of the four Ogoni chiefs by a mob points to a profound shaping of consciousness by Nigeria's political history that Saro-Wiwa did not recognize. Quayson perceptively argues that the root cause of these murders was the culture of violence, distrust, and impunity instilled throughout Nigeria by corrupt, vicious military regimes. In Nigeria, those in positions of power were so regularly bribed into supporting the status quo that to be perceived as speaking out in favor of the official position (as were the four chiefs) was to be marked as a traitor. In turn, the lesson instilled by Nigerian governance and society was that violence was the means with which to deal with those deemed to oppose you; since those motivated by Saro-Wiwa "began to articulate their perception of political action partially from within the ethical framework provided by the system of patronage and violence," in his absence they "deployed the means of understanding politics that was discursively instituted in Nigerian politics itself" (73).

The very notion of an Ogoni ethnic identity has also been frequently questioned since Saro-Wiwa's death. As Michael Watts notes, "There was no simple Ogoni 'we,' no unproblematic unity, and no singular form of political subject" ("Petro-Violence" 211). There are five subgroups in Ogoni; each subgroup has a common myth of origin, while the Ogoni as a whole do not, and each speaks "a mutually unintelligible language" (68). As a result, these subgroups "tend to engender stronger loyalties than the Ogoni 'nation'" (Osaghae 328). In other words, scholars such as Watts and Eghosa Osaghae are suggesting that a sense of pan-Ogoni identity was not necessarily something predating colonialism and postindependence Nigeria, but something that Saro-Wiwa constructed.

According to Watts, this construction has had profoundly problematic effects. He situates Saro-Wiwa's "myth of a pan-Ogoni identity" in relation to the creation "of governable (or un-governable) spaces" by "the baneful twins of authoritarian government and petro-capitalism" ("Violent" 282). This myth, he argues, cannot be separated from "spoils politics in Nigeria," in which "ethnic claims-making" plays a significant role in competition for oil wealth and the multiplication of massively corrupt subnational "local government areas," and from a national imaginary in which the nation is viewed as "simply a big pocket of oil monies to be raided in the name of indigeniety" (291–92). In ignoring or downplaying the divisions among the Ogoni (which cannot be separated from the politics of oil wealth), Saro-Wiwa helped create an indigenous subject and an indigenous space that "was contentious and problematic" and that after his death "unraveled into fragments of class, clan, generation and gender" and resulted in the decline of MOSOP (289). In focusing on "the need for ethic minority inclusion as the basis for federalism," Saro-Wiwa ignored "the histories and geographies of conflict and struggle among and between ethnic minorities" shaped by petro-capitalism (290), and to some degree, that aspect of his narrative ended up working within (rather than against) the discourse driving the "intense conflict" for oil wealth "as more oil-producing minorities (for instance, the Adoni, the Itsekiri, the Ijaw) clamor for more compensation and for the recognition of their claims for resource control" (279).

More generally, events since Saro-Wiwa's death have pointed toward the problems with MOSOP's focus on compensation and with the notion that a semiautonomous Ogoni state would necessarily have entailed better management of petrodollars. A significant component of the Ogoni Bill of Rights focused on the "oil royalties and mining rents amounting to an estimated 20 billion US dollars" MOSOP claimed was owed to the Ogoni (Saro-Wiwa, *Genocide* 98), and Saro-Wiwa called for "the percentage of oil revenue" paid to "the minorities of the delta and its environs" to be increased "radically" (*Month* 45). The obvious question is: Would a semiautonomous Ogoni state flush with petrodollars from reparations and from increases in percentage of oil revenues have been able to escape the corruption, division, violence, and injustice that have accompanied petro-capitalism more generally in Nigeria? Recent history suggests the answer is probably not. Since democratic elections brought Obasanjo to power, the allocation of derivation income to the Niger Delta has increased dramatically: "Ken Saro-Wiwa could plausibly claim that the oil-producing states and the local councils were starved of oil revenues . . . but that situation has changed" (Watts, "Sweet" 46). In "the new fiscal environment," corruption has been "decentralized," "the means of violence" democratized, and a new group of local "machine politicians" has arisen; from "this trio of forces" a new "wave of violence . . . emerged," including intercommunity conflicts, urban interethnic warfare, intracommunity youth violence, and "extraordinary state violence" by federal security forces entering the fray and supposedly seeking out militants (46).

Throughout his writing, Saro-Wiwa clearly endorses Fanon's insight in *The Wretched of the Earth* that the "battle against colonialism does not run straight away along the lines of nationalism" (148)—that is, that the goal of national independence was not necessarily identical with the goal of dismantling an unequal, exploitative colonial system and culture because these could live on, albeit in new forms, with new players in old roles. What he did not, perhaps, consider fully enough was that the goal of increased ethnic federalism (increased rule of Ogoni by Ogoni) is not necessarily identical with the goal of liberation from imperial development in Nigeria and its dreadful effects.

Reviving Resistance: Chinua Achebe

Although *Arrow of God* was written almost thirty years earlier than *Genocide in Nigeria* and it focuses on a moment well before the discovery of oil in the Niger Delta, Achebe constructs a counternarrative of development strikingly similar to Saro-Wiwa's. Both authors emphasize how sacrifices encouraged by imperialism in the name of progress led to degeneration and disaster. In fact, examining *Arrow of God* in relation to *Genocide in Nigeria* and *A Month and a Day* highlights how very prescient Achebe was in terms of anticipating how colonial development would lead to violence, corruption, division, and arrested decolonization in the decades following independence. Furthermore, like Saro-Wiwa, Achebe suggests that reversing the catastrophe of Nigeria's historical trajectory necessitates decolonizing the valuation of nature and indigenous culture and reviving a precolonial culture's reverence for and close links with a local ecological community.

Of course, in many ways *Genocide in Nigeria* and *A Month and a Day* offer a more current, more activist-oriented anticolonial environmentalist vision than does *Arrow of God*. Most specifically, Achebe does not echo Saro-Wiwa's particular concern with how microminority groups like the Ogoni and their environments are devalued, instrumentalized, and decimated in contemporary Nigeria. More generally, Saro-Wiwa's focus on the imperial effects of increasingly powerful, unregulated transnational capital in the last quarter of the twentieth century results in an emphasis on economic motivations and structural relationships relatively absent in *Arrow of God*. Saro-Wiwa also foregrounds in ways that Achebe does not how an alternative kind of development, both personal and collective, can be forged in active struggle. Finally, in that Saro-Wiwa is substantially more focused on the actual dangers of ecological damage than Achebe, he is the more obviously environmental writer.

Yet putting Saro-Wiwa's writing in dialogue with *Arrow of God* not only helps us think about how Achebe's novel might be made relevant to the present; it also points to ways that *Arrow of God* might contribute to a reconsideration of the narratives of identity embedded in *Genocide in Nigeria* and *A Month and a Day*. Saro-Wiwa suggests that accessing a historically transcendent Ogoni identity is MOSOP's key to decolonized consciousness and to a proper ecological sensibility. Once it is uncovered, this identity becomes the foundation both for resistance and for a semiautonomous Ogoni state liberated from the effects of (neo)colonial development. He even suggests that such a state would necessarily be ecologically sensitive since MOSOP's anticolonial struggle is grounded in an Ogoni "genius" in tune with the spirit of local nature. The potentially problematic implications of this narrative are highlighted when we read it in relation to *Arrow of God*. Achebe suggests that grounding resistance in the notion of a transcendent collective identity characterized by a sameness of outlook ultimately suppresses the differences that must be negotiated and hampers the work necessary to forge unity in diversity. More generally, he emphasizes the transformation over time of social and ecological conditions in ways that bring into question the possibility of transhistorical identity. In *Arrow of God*, effective cultural practice is based on a principle of extravagant aberration and transformation rather than return. In fact, *Arrow of God* would suggest that Saro-Wiwa is hewing too closely to a colonial vision when he endorses the notion of internally unifying ethnic essences that create permanent, clear boundaries among peoples (and nations) and, as a result, hamper recognition of internal divisions.

Focused on the period following "the pacification of the lower Niger" depicted in *Things Fall Apart*, *Arrow of God* foregrounds British attempts to consolidate control through centralized, widespread socio-political engineering and through the creation of Nigeria as a new co-lonial geopolitical unit. The colonial administration is busily trying to destroy or assimilate Igbo social and cultural authority and bringing the few remaining resistant communities into its Nigerian order. If im-perial discourse would represent this process as progress, as the move-ment from chaos toward order, *Arrow of God* depicts it as generating violence, division, injustice, corruption, and loss of self-determination. Ultimately, the novel's central tragedy is not the fall of an individual, Ezeulu, but the slow, ongoing deterioration of Igbo cultural principles and the formation of consciousness and of an incipient nation along colonial lines. If one reads the novel in terms of its moment of produc-tion, Achebe is suggesting that this formation has set up the disastrous situation developing in the postindependence nation during the 1960s. However, he also suggests there is hope for an alternative trajectory of development if Igbo philosophical principles can be rejuvenated and embraced. This hope does not entail a return to social models, politi-cal identities, cultural institutions, or ecological relationships unmedi-ated by colonial history. In *Arrow of God,* society and nature are always historically situated, and history cannot be transcended or reversed. In fact, the damage caused by colonialism includes its undermining of a cultural orientation based on awareness of a fluid, heterogeneous reality refusing assured knowledge and control. For Achebe, an alternative tra-jectory of development will be based on aberrant identities and institu-tions emerging out of existing conditions and on skepticism regarding the notions of transcendence and mastery encouraged by colonialism. It will also be associated not only with an ecological awareness of interde-pendence with the nonhuman natural world but also with a sense that nature and society are historically and dialectically bound.

Things Fall Apart famously ends with a colonial administrator re-flecting on the ways he will fit Okonkwo's story into a paragraph or two in the book *The Pacification of the Primitive Tribes of the Lower Niger.* This title emphasizes a much larger and more disastrous discursive vio-lence: the suppression of a rich precolonial history and culture through the reductive classification of colonial epistemology. Early on in *Arrow of God,* Achebe includes a scene in which Tony Clarke, a junior officer, reads the concluding chapter of the administrator's fictional book. If, on the whole, Clarke considers the narrative from the earlier colonial era "smug," he does appreciate its conclusion, which sums up the process of

colonial bureaucratic development—of law giving, economic rational-
ization, infrastructure building—in heroic terms: that is, the way it creates for him, who has been shut out from the first stage of conquest, a
still-pioneering role:

> For those seeking but a comfortable living and a quiet occupation
> Nigeria is closed and will be closed until the earth has lost some
> of its deadly fertility and until the people live under something
> like sanitary conditions. But for those in search of a strenuous
> life, for those who can deal with men as others deal with material,
> who can grasp great situations, coax events, shape destinies and
> ride on the crest of the wave of time Nigeria is holding out her
> hands. For the men who in India have made the Briton the law-
> maker, the organizer, the engineer of the world this new, old
> land has great rewards and honourable work. I know we can find
> the men. Our mothers do not draw us with nervous grip back
> to the fireside of boyhood, back into the home circle, back to the
> purposeless sports of middle life; it is our greatest pride that
> they do—albeit tearfully—send us fearless and erect, to lead the
> backward races into line. "Surely we are the people!" (33)

The image of Africa as the embodiment of a "deadly" and powerfully
fertile wilderness designates the continent as the ultimate test of the colonizer's ability to master nature through rational organization and engineering; such mastery is the means by which the British will establish
once and for all their transcendent place in "the game of life." Yet the
metaphor of "the game" already brings peoples and places "into line"
with a particular hierarchical organization based on the categories of
nature and culture. Designating the "backward races" (who need to be
dealt with as "material") as part of the natural world, the game can only
establish the differences among European colonizers. More generally,
playing the game (engaging in colonial development) entails accepting a
disciplinary knowledge that constructs a wilderness to be transformed
by the winner.

Throughout *Arrow of God,* Achebe mocks such naturalization of
authority by emphasizing its contradictions, its absurd misrepresenta-
tions, and its alienating effects. Listening to "the distant throb of drums"
while lying awake at night, Captain Winterbottom wonders "what un-
speakable rites went on in the forest at night, or was it the heart-beat of
the African darkness?" (30). Winterbottom believes he "understands"

the Igbo because of his long experience in Africa; however, through the depiction of a rich Igbo cultural life, the novel sets up these Conradian clichés as ridiculous. "One night" Winterbottom has an inkling that what he hears could be "throbbing . . . from his own heat stricken brain" (30). This thought is richly evocative, suggesting a relationship between a cultural narrative (the heart of darkness) and a fever-induced delusion. However, insight is made impossible by the ease with which his thoughts are brought back in line with colonialism's map; the problem is in the place: "this dear old land of waking nightmares!" (30). The novel has many such moments when absolute confidence in colonial classification suppresses interrogation and generates alienation. This confidence, ironically, prevents engagement with what in the novel is a fluid, diverse reality resisting reduction to clearly bounded categories and, more generally, to what Achebe refers to as a malignant fiction: a single, universalizing authoritative narrative suppressing anything it does not already admit (Achebe, *Hopes* 138–53).

In many ways, Saro-Wiwa's representations of colonial consciousness echo Achebe's; in particular, *Genocide in Nigeria* suggests how the categorization of peoples based on their closeness with a dangerous, wild African nature (and corresponding distance from civilization) was closely linked with the disempowering of the Ogoni and with their instrumentalization in colonial and then postcolonial Nigeria. Saro-Wiwa also, like Achebe, emphasizes the beneficial qualities in an indigenous African culture that get suppressed by the myopic vision instilled by colonial consciousness and that are sacrificed in the name of development.

In his essays, Achebe often contrasts colonialism's classificatory discourse with Igbo philosophy and cosmology. He suggests that Igbo epistemology is both historical and relational (projecting identities as fluid, interconnected, and multiple) and that it encourages skepticism regarding authoritative explanatory frames. Achebe has famously described the Igbo as embracing "a dynamic world of movement and of flux" and proclaimed that the "Igbo formulate their view of the world as: 'No condition is permanent'" (*Hopes* 62, 64). In turn, an awareness that meaning and identity are in movement and depend on historical conditions encourages adaptability and a willingness to question what is already known. One must be open to an otherness that will reveal the limitations of any single perspective or narrative (no matter how old or widely accepted). According to Achebe, the Igbo embrace "the middle ground," which is "the home of doubt and indecision, of suspension of disbelief"

(*Education* 6). As a result, he proclaims, Igbo art mediates "between old and new, between accepted norms and extravagant aberrations. Art must interpret all human experience, for anything against which the door is barred can cause trouble. Even if harmony is not achievable in the heterogeneity of human experience, the dangers of an open rupture are greatly lessened by giving everyone his due in the same forum of social and cultural surveillance" (*Hopes* 65). In *Arrow of God,* this view is summed up in the proverb proclaiming that "in a great man's household there must be people who follow all kinds of strange ways . . . in such a place, whatever music you beat on your drum there is somebody who can dance to it" (46). Achebe is militantly particularist not only in the sense that he questions universalizing colonial discourse by giving an opposing perspective from a particular (marginalized) position but also in the sense that he embraces a philosophy that denies the validity of all universalizing discourses.

In addition, Achebe suggests that Igbo philosophy represents subjectivity as mediated and decentered. If the Igbo "postulate an unprecedented uniqueness for the individual," they also see "him" as "in a very real sense subordinate to his community" and "subject to the sway of non-human forces in the universe" (*Hopes* 57). In this view, the individual is neither autonomous nor determined; instead, she is constantly engaged in processes of mutual determination with what is "outside." As a result, one can neither transcend the perspective shaped by one's relationships, nor is one's vision completely determined by them. This outlook, Achebe claims, results in cultural forms that encourage a democratic ethos working against centralized power, extremism, and notions of transcendence. The Igbo, Achebe proclaims, balance an "unsurpassed individuality, by setting limits to its expression. The first limit is the democratic one, which subordinates the person to the group in practical, social matters. And the other is a moral taboo on excess, which sets a limit to personal ambition" (58).

In *Arrow of God,* Umuaro is grounded in the kind of Igbo worldview Achebe describes in his essays. Before the time when "the hired soldiers of Abam used to strike in the dead of night, set fire to the houses and carry men, women and children into slavery," the six villages that make up Umuaro "lived as different people and each worshipped its own deity." However, "to save themselves" they reformulated their identities and their cosmology by hiring "a strong team of medicine-men to install a common deity" called Ulu and by creating a new, larger community: "From that day they were never again beaten by an enemy" (14–15). Faced with a crisis for which existing sociopolitical identities are inadequate, the villages demonstrate a profound adaptability. They do not give up their existence as social units, but they do change their positions in relation to one another and create something new, an "extravagant aberration" from what already exists. This process is based on the prioritizing of communal survival and of interdependent connection. Umuaro is also shaped in such a way as to limit the power of any one village and to curtail ambition: "when the six villages first came together they offered the priesthood of Ulu to the weakest among them to ensure that none in the alliance became too powerful" (15). The fact that "from that day they were never beaten by an enemy" suggests that Umuaro's foundational principles are powerful means of resisting external control and fighting for communal survival and prosperity.

In that Saro-Wiwa sought a transformed kind of nation-state, his call for a federated Nigeria seemed to encourage the kind of "extravagant aberration" emblematized by Umuaro in *Arrow of God*. This goal was expressed in the Ogoni Bill of Rights, which was presented to the Nigerian government in 1990 and called for the Ogoni to "be granted Political Autonomy to participate in the affairs of the Republic as a distinct and separate unit" (*Month* 48). Saro-Wiwa believed that such autonomy was a means to bring about improved civil liberties and rights, economic and cultural development, better environmental protection, and a more democratic Nigeria. In fact, he saw the struggle for ethnic autonomy within the nation as the hope for a decolonized and democratic Africa. He claimed it would be "wonderful" for Africa "if the various ethnic nations that make it up could assert themselves" in ways similar to the Ogoni: "We would be heading for a more democratic system far from the dictatorships which have ruined the continent; and we might succeed in reordering our societies, so that there would not be so much exploitation" (*Month* 92). However, although they both seem to embrace a model of development rooted in the transformation of political institutions and in aspects of precolonial culture, Saro-Wiwa's underlying approach to ethnicity is profoundly at odds with Achebe's embrace of flexible (relational) identities. In contrast with Saro-Wiwa, Achebe does not understand collective identity in bounded, transhistorical terms and does not ground development in a return to identities from the past. For him, any movement forward depends on a more substantial process of extravagant aberration.

In his essays, Achebe also points to the kind of ecological philosophy and ethos entailed by his Igbo "world." He stresses the ways that Igbo culture ascribes powerful spiritual and moral agency to local nature, claiming that "the earth goddess, Ala or Ana," is in "the Igbo pantheon" both "fountain of creativity in the world and custodian of the moral order in human society. An abominable act is called *nso-ana,* 'taboo to earth'" (*Education* 108). Such a cosmology projects reverence and a desire for connection with (rather than transcendence of) the natural world on which one depends for survival and guidance. At the same time, Achebe suggests that the Igbo view the natural world as historical. He claims:

> In Igbo cosmology even gods could fall out of use; and new forces
> are liable to appear without warning in the temporal and meta-
> physical firmament. The practical purpose of art is to channel
> a spiritual force into an aesthetically satisfying physical form that
> captures the presumed attributes of that force. It stands to reason,
> therefore, that new forms must stand ready to be called into being
> as often as new (threatening) forces appear on the scene. It is
> like "earthing" an electrical change to ensure communal safety.
> (*Hopes* 64)

Given the animistic aspect of Achebe's "Igbo cosmology" and, more generally, given the central role of local nature in the Igbo societies represented in his novels, this statement suggests a natural world that changes in profound ways; it too is subject to the dialectic "between old and new, between accepted norms and extravagant aberrations." In fact, the guiding role Achebe ascribes to "the earth" for Igbo culture would suggest that the flexibility of that culture has been shaped by the flu- idity of the natural world; at the same time, the notion of channeling "new (threatening) forces" suggests a shaping of nature by Igbo culture. In this sense, Achebe offers a more historical and dynamic vision of nature than does Saro-Wiwa. If the picture of precolonial Ogoniland at the beginning of *Genocide in Nigeria* is a still life of an unchanging natural environment and culture, in Achebe's writing the precolonial Igbo world "is like a Mask dancing. If you want to see it well you do not stand in one place" (*Arrow* 46).

In *Arrow of God,* the conception of nature entailed by Achebe's phi- losophy is apparent in the kind of dialogic attentiveness to natural phe- nomena entailed in cultural practice. If "no condition is permanent," then one must be constantly on the lookout for changes that slip out of existing narratives and challenge expectations. For the sake of survival and prosperity, one must not only try to listen to the different voices of nature just as one listen to the different voices of the human community but also be vigilant not to reduce the meanings of those voices to an already existing explanatory system when they do not fit. This particular kind of attentiveness is embedded in Ezeulu's rituals as he watches "for signs of the new moon" at the beginning of *Arrow of God*. Although "he knew it would come today," he still "always began his watch three days early because he must not take a risk" (1). Ezeulu is extremely self-confident regarding his knowledge and does not like to consider the possibility that he could be wrong. However, his cultural role necessitates that he watches, and the watching has made him cautious. In particular, "when the rains came" and "the new moon sometimes hid itself for days behind rain clouds," he would "peer and search" every evening "while it played its game" (1). For the colonists, the "game of life" is played on the board (so to speak) of an objectified natural world, but for the Igbo an animated nature is a player refusing knowledge of its next move. Ezeulu has even constructed his hut in such a way that it will allow better observation of the sky, and as he watches "he constantly blinked to clear his eyes of the water that formed from gazing so intently" (1). Normally, he does not like "to think that his sight was no longer as good as it used to be," and this reluctance to admit the possibility of a change in his powers of observation and consequently in his position anticipates the problems that will be brought about by his pride; nonetheless, when he sees the very "thin" moon, "he peered more closely to make sure he was not deceived by a feather of cloud" (1–2). In this sense, he recognizes the possibility of error in his observation as a result of physical limitations and tries to guard against it.

In contrast with Igbo cosmology, colonial Christianity in *Arrow of God* objectifies local nature and discourages a sense of interdependence with it. The teacher Mr. Goodcountry proclaims to the converts, "You must be ready to kill the python as the people of the rivers killed the iguana. You address the python as Father. It is nothing but a snake, the snake that deceived our first mother, Eve. If you are afraid to kill it do not count yourself a Christian" (47). Goodcountry wants to purge local nature of any benevolent authority and significance, and he wants to sever the converts' sense of familial or communal reliance on it. Toward this goal, he reduces the python's meaning to the predatory threat associated with the more general category of "snake" and to the position of the snake in an abstract biblical narrative generated elsewhere. In contrast, the people of Umuaro recognize that pythons are relatively benevolent and even helpful ("they did no harm and kept the rats away") and can be safely allowed to live in peoples' huts (50). These people are not represented as having escaped the mediating role of culture and language; for example, they still use the more abstract category of "ordinary snake": "when a man sees a snake all by himself he may wonder whether it is an ordinary snake or the untouchable python" (143). However, in contrast with Goodcountry's story, even this proverb implies a narrative regarding a plurality in local nature (which is both benevolent and threatening) and a need for attention to conditions in the local natural world.

The implications of Goodcountry's views are reflected in the action he inspires. As a result of the sermon, Ezeulu's convert son Oduche locks the royal python in a box; when caught, he is accused of "desecrating the land" (131). Given the symbolic significance of the python in Umuaro, this accusation makes sense in various ways: Oduche has, in fact, stripped the land of some of its spiritual value (made it less sacred), the respect it is normally accorded, and, not least, its value as a part of the community. More generally, the abstract Christian narrative of salvation through transcendence of nature threatens to undermine the Igbo emphasis on survival and the well-being of the individual and community through reliance on the earth. When Ezeulu learns that the church bell "is saying: Leave your yam, leave your cocoyam and come to church," he proclaims, "then it is singing the song of extermination" (43).

The lack of respect for animistic religious belief expressed by Oduche's transgression and the divisiveness it causes is part of a larger pattern of conflict and "crisis of authority which colonialism triggers in Igbo culture" (Gikandi 69). Through selective use of force and the generation of divisions among and within groups, the British have been able to "pacify" all of what is now Nigeria and are in the midst of a process of institutional engineering intended to develop the region in ways that will bring it in line with colonial principles, interests, and centralized control. Umuaro has resisted cultural and social assimilation longer than the surrounding communities; however, as traditional authority and institutions are put under pressure, Umuaro is increasingly divided and in chaos.

In *Arrow of God*, change—even substantial change—is not the problem; the founding of Umuaro, for example, suggests that a reformulation of identity and cultural institutions can be a kind of progress. The long-term threat is the particular kind of change encouraged by the formation of Nigeria, which pushes individuals and communities to follow and, increasingly, internalize colonial principles as they struggle to survive and prosper. According to Simon Gikandi, under the new dispensation "the only sanctioned cultural text—and social fiction—is that backed by the power of the colonial state" (61). As a result, the kind of cultural principles on which Umuaro was founded and that the novel suggests would be crucial for the formation of effective resistant identities and institutions are being weakened; Umuaro, and by implication all of Nigeria, is being disciplined in such a way that resisting colonial development becomes harder and harder. Although Saro-Wiwa explicitly invokes a longer colonial and postcolonial history than does Achebe, both foreground how the trajectory initiated by colonialism is a disaster in terms of growing corruption, communal conflict, instrumentalization, and external manipulation.

In *Arrow of God*, British identity and culture have become associated with a transcendent authority to which people try to attach themselves and to which they aspire. As Akuebue notes, "the masked spirit of our day is the white man and his messengers" (154). At the same time, as Akuebue also makes clear, the authority of the British is never matched by their servants: "I have never heard of a messenger choosing the message he will carry. Go and tell the white man what Ezeulu says. Or are you the white man yourself?" (140). By creating a situation in which those working with them gain some authority but in which the colonized can never achieve the status of the colonizer (always not quite white), the British both encourage their colonial subjects to do their bidding and accord themselves a transcendence out of their colonial subjects' reach. Certainly, as Homi Bhabha suggests, the contradictions in this model suggest that it may contain in itself the seeds of its own undoing; however, *Arrow of God* still emphasizes that it was a powerful means of creating willing servants and dividing the colonized. At the collective level, subjection is fostered by privileging communities that are more culturally accommodating to the British and that accept their authority. When war broke out between Okperi (home to a British government station) and Umuaro (which resisted British control) over some land, Winterbottom settled the dispute by giving it to Okperi. In telling the story, Winterbottom proclaims, "Okperi welcomed missionaries and government while Umuaro, on the other hand, has remained backward" (36).[5]

In *Arrow of God,* colonial capitalism is quickly becoming the most effective means of disciplining consciousness and identity. In articulating the perceived dangers of falling behind in "the race for the white man's money," Nwabueze explains to Ezeulu and Akuebue that Umuaro loses power as "people from other places are gathering much wealth" and "control the great new market": "We have no share in the market; we have no share in the white man's office; we have no share anywhere" (169–70). As the competition among communities becomes ever sharper, boundaries harden, and communal identities calcify. In turn, because this competition is primarily defined in monetary terms, the colonized grow reliant on an abstract system of finance and trade over which they have little control. Just as important, the focus on money encourages increasing instrumentalization—including of the self. For example, the "paid gang" working on the road "were as loyal as pet dogs," according to the British overseer, and are in sharp contrast to the unpaid "free but undisciplined crowd from Umuaro" (76–78).

The colonizer's approach to manipulating and distributing power in Nigeria exponentially increases its abuse. While undermining existing cultural constraints, the British encourage an understanding of authority that divorces it from responsibility or responsiveness to the communities over which one has control. The result is a growing institutionalizion of corruption. The native police, who come from communities outside Umuaro, use their power to get "kola" from Ezeulu's family through blackmail (154). Such systematic, petty corruption becomes more severe when district officers like Winterbottom, working among "tribes" like the Igbo "who lack Natural Rulers," are instructed to appoint "warrant chiefs" in order to develop "an effective system of 'indirect rule' based on native institutions" (56–57). The British rationalize this means of control by claiming that it can help "build a higher civilization upon the soundly rooted native stock that had its foundation in the hearts and minds and thoughts of the people." This rationale becomes another means for Achebe to expose the discourse of natural development as a colonial fantasy driven by the desire for mastery. The Igbo have no "native institutions" based on centralized, nondemocratic authority, and rather than creating a "higher civilization," the British destroy a "native system" that effectively places limits on power and encourages respect for otherness (and that has its "foundation" in the people). In their arrogance and drive for control, the British create not "higher standards," but corruption and cruelty. The man Winterbottom appoints sets "up an illegal court and a private prison," and when Winterbottom suspends him "the Senior Resident who . . . had no first-hand knowledge of the matter ruled that the rascal be reinstated. And no sooner was he back in power than he organized a vast system of mass extortion" (57–58).

Winterbottom sets down the reinstatement to bureaucratic distance from local reality and to the structure of "British colonial administration." If, to some degree, he is correct, the novel also mocks his belief that he is "the man on the spot," with the requisite knowledge to make informed decisions (57). He too is distanced from the reality around him by his colonial authority and attitudes. In addition to his original appointment of the chief, the "programme of road and drainage construction following a smallpox epidemic" organized by Winterbottom enabled the chief to extort money through blackmail and to destroy the homes of those who did not pay up (58). Winterbottom may have tried to ensure that the plans "did not interfere with people's homesteads," but his own ignorance led to disaster. In turn, his arrogant confidence in colonial identities enables him to avoid self-reflection and critical analysis of the underlying causes for the "system of extortion": that is, colonial development and the principles underpinning it. He attributes the episode to the "elemental cruelty in the psychological make up of the native that the starry-eyed European found so difficult to understand" (58).

Among the most dangerous aspects of Nigeria's colonial formation is such hampering of skepticism and self-scrutiny. The colonizers' blind confidence and profound inability to understand the places they rule contribute to the formation of policies with destructive effects they do not anticipate. At the same time, their lack of respect for alternative points of view and inability to doubt their own perspective leave them unable to reflect effectively on the historical processes they set in motion. *Arrow of God* emphasizes that these processes are not the result of their ability to understand and control conditions; the novel emphatically denies the British (and mocks their belief in) such mastery. However, it does suggest that their naturalizing representational economy enables them to avoid responsibility and to suppress their own limitations. At the same time, as the colonized are themselves interpolated as colonial subjects, they too increasingly lose the ability to question themselves and the assumptions inculcated by colonial development, as well as to confront their own responsibility for deteriorating conditions. Hampering creativity, colonial development results not only in ever more conflict, corruption, and chaos but also in an increasing inability to contemplate or initiate a national historical trajectory different from the one Saro-Wiwa traces in *Genocide in Nigeria* and *A Month and a Day.*

The ending of *Arrow of God* also offers a warning against malignant fictions that create the illusion of mastery, shift responsibility, and suppress healthy skepticism; however, in this case these fictions are associated not only with colonialism but also with the narratives of identity and sacrifice Ezeulu and Umuaro draw on in their conflict with one another. (In this sense, Achebe frustrates any disciplining of the story he tells using easy assignations of blame and innocence.) Ezeulu is usually confident that he knows the will of Ulu when the rest of Umuaro does not and that he has full understanding of events and their causes. Such an attitude hampers the humility needed to reflect critically on his own contribution to the crisis of authority and the divisions in Umuaro. In a conversation focused on responsibility for the strengthening of colonialism's grip on Umuaro, Akuebue tells Ezeulu that much of the community blames him for agreeing with Winterbottom in the dispute about the land in Okperi and for sending his son for a Christian education. In response, Ezeulu places responsibility entirely on the rest of Umuaro and offers this message to the community: "The man who brings ant infested faggots into his hut should not grumble when lizards begin to pay him a visit" (131–32). *Arrow of God* is rife with such proverbs regarding the need to accept responsibility, but in this crisis both Ezeulu *and* the citizens of Umuaro are all too ready to point the finger elsewhere.

In this scene, Achebe also suggests how a malignant fiction of sacrifice and progress all too easily results in unacknowledged destructive consequences and the externalizing of responsibility. In response to the accusation that he betrayed Umuaro by sending his son to the missionaries, Ezeulu claims that it was a means to gain knowledge and power in order to fight the British. Oduche, he argues, is a necessary sacrifice: "A disease that has never been seen before cannot be cured with everyday herbs. When we want to make a charm we look for the animal whose blood can match its power; if a chicken cannot do it we look for a goat or a ram; if that is not sufficient we send for a bull. But sometimes even a bull does not suffice, then we must look for a human" (133). Akuebue interrogates the wisdom of this decision by asking Ezeulu if he really knows what he is sacrificing; in response, Ezeulu claims that, as Oduche's father and as Ulu's priest, he understands the identity of what is sacrificed and the outcome of the sacrifice. However, Oduche cannot be instrumentalized and will not be the arrow in Ezeulu's bow. He accepts the principles of Christianity and turns away from Ezeulu's authority and Umuaro's traditions.

At the ending of *Arrow of God*, Umuaro positions Ezeulu himself as a kind of sacrifice. They claim the meaning of his downfall and death is "simple": "Their god had taken sides with them against his headstrong and ambitious priest and thus upheld the wisdom of their ancestors—that no man however great was greater than his people; that no one ever won judgement against his clan" (230). If this narrative reiterates some key principles underpinning the founding of Umuaro (the grounding and circumscribing of power in community), its reductive simplicity also prevents inquiry into the complex causes of Ezeulu's downfall and the threat to the community. By placing blame on the chief priest, it enables the community to avoid reflecting on what it has done and is doing to endanger itself. In the past, the priest of Ulu functioned as a kind of scapegoat meant to take on "the sins of Umuaro" in order for the community to be cleansed and survive—as is demonstrated by his role in the Festival of Pumpkin Leaves (73). However, if the position of Ezeulu in Umuaro's discourse at the end foregrounds "the scapegoat theme," which "was widespread in West African writing during the period of the publication of *Arrow of God*," it is not necessarily clear that, as Dan Izevbaye argues, Ezeulu fulfills the role of scapegoat by saving his community (37). Rather, by alluding to ongoing threats at the end to the community's well-being and to the dangers of the community's rhetoric of blame, Achebe seems to question the very discourse of sacrifice embodied in the scapegoat.[6]

Achebe subtly signals the dangers of Umuaro's narrative in the final paragraph. The narrator writes of the vast increase in converts resulting from the promise that those sacrificing yams to the Christian god would be protected from Ulu's anger: "The Christian harvest . . . saw more people than even Goodcountry could have dreamed. In his extremity many a man sent his son with a yam or two to offer to the new religion and to bring back the promised immunity. Thereafter any yam harvested in his fields was harvested in the name of the son" (230). In looking for protection ("immunity") from a delocalized, transcendent god, the people of Umuaro do not uphold "the wisdom of their ancestors." They give up responsibility for their own survival and severely attenuate the sense that strength is to be found in communal unity and connection with the land. It may be that in turning toward Christianity and away from Ulu they embrace an Igbo cultural practice of destroying and creating gods in order to protect the community in the face of new historical threats; however, in making this move they risk letting go of the very principles underpinning that practice, since Christianity teaches that there is only ever one true god who transcends place and time and that all other gods are false.

The likelihood that Umuaro will accept such a doctrine is reflected in the way the community has already separated Ulu from their own agency. Using an ambiguous indirect discourse as he does throughout the narrative, Achebe suggests that Umuaro's "simple" story of Ezeulu's downfall sets the blame for Ulu's incipient downfall on the god: "in destroying his priest he had also brought disaster on himself, like the lizard in the fable who ruined his mother's funeral by his own hand. For a deity who chose a moment such as this to chastise his priest or abandon him before his enemies was inciting people to take liberties" (230). In this narrative, Ulu is divorced from the will of the community that brought him into being. In this sense, it mirrors a perspective that colonial Christianity will only further encourage: a belief in an abstract transcendent spiritual realm and in representations of identity and causation that externalize responsibility and reduce human and ecological agency.

If hope for a decolonized Nigeria in *Arrow of God* lies in the generation of resistant sociopolitical identities and cultural practices through the principles used to make Umuaro, the most depressing aspect of the ending is its suggestion that those principles will increasingly be sacrificed as Nigeria is disciplined by colonial development. In the creation of Umuaro, the six villages responded to an external threat by creating a new identity founded on the principles of unity in diversity, the just distribution and limitation of authority, openness to debate among differing points of view, and power rooted in responsibility and responsiveness to the interdependent socioecological community constituting place. They embraced the idea that protection and progress required both acknowledging that reality cannot be contained by "accepted norms" and a willingness to revise identities and temporal trajectories.

Such a perspective generates "extravagant aberrations" and the skepticism required for still further revision.

In contrast with Igbo culture and society, *Arrow of God* suggests, the development of Nigeria along colonial lines encourages a belief in strictly bounded, ahistorical identities; competition rather than cooperation; the pursuit of the transcendent (unchecked) power of the colonizer; and a lack of skepticism regarding accepted, authoritative narratives. In this sense, Achebe grounds the dysfunctional condition of the postindependence nation both in colonial history and in the failure among Nigerians to identify and challenge (neo)colonial forms of consciousness among themselves in the present. They have been unable to develop a form of collective, national identity that would represent an "extravagant aberration": that is, one based on a unity of interdependent older communities coming together for mutual protection and strength and on the grounding of authority in identification with the new nation in all its diversity. The result of this failure is a corrupt ruling elite focused on taking over the roles and status of the colonizer, unscrupulous competition for resources among ethnic communities, the growth of unjust inequalities within the nation, and increasing conflict. In these conditions, external manipulation and exploitation thrive; as one proverb in *Arrow of God* suggests, when "brothers fight a stranger reaps the harvest" (131).

Among the long-term serious consequences of colonial development in *Arrow of God* is the transformation of the relationship with the land. An early example in the novel involves the disputed area between Okperi and Umuaro. When, in his judgment of the case, Winterbottom determines it to be owned by Okperi beyond "any shade of doubt," he strips the place of a rich identity determined by a subtle history and complex pattern of use (37). The judgment offers a particularly vivid example of the process of deterretorialization and reterritorialization entailed by colonial modernity.

At the end of the novel, the offering of yams to an abstract, universal God rather than to a god embodying place signals a subtle but substantial change among the citizens of Umuaro in the conception of the natural world and of humans' relationship with it. No longer will survival and prosperity be viewed as dependent on local nature, understood as filled with powerful subjectivity and as deserving reverence. Instead, Umuaro will increasingly accept colonial Christianity's construction of progress as mastery of nature, which becomes dead material and/

or the embodiment of evil to be destroyed or utterly transformed. As Goodcountry says of the python when urging its destruction, "It's only a snake." *Arrow of God* indicates that, as they are disciplined by colonialism, Nigerians increasingly embrace a reductive, instrumental view of local nature's meaning and will sacrifice it and their prior relationships with it in the pursuit of the magical powers of money, the Christian god, and the other trappings of European culture. In turn, the threat to self-determination and survival grows; for example, communities become increasingly dependent on an abstract monetary system, on commodities, and on the market. In this sense, what we would now call an ecological sensibility is important for understanding the wider implications of Achebe's perspective on the destructive consequences of sacrifice driven by malignant fictions of development. In accepting the inferior otherness of what is associated with nature and its consequent sacrifice in the name of progress, Umuaro brings disaster on itself, "like the lizard in the fable who ruined his mother's funeral by his own hand" (230).

Saro-Wiwa clearly echoes Achebe's concern with the objectification and sacrifice entailed by a colonial discourse of development. He highlights the devastation this discourse has brought not only to the Ogoni and the ecosystem of the Niger Delta but also to all of Nigeria and Africa. Like Achebe, he also points to the ways this discourse leads the colonized to sacrifice cultural principles that might enable them to fight their own subjection. In a sense, his mobilization of MOSOP might even be aligned with Achebe's model of transformed identity; his efforts to foster a sense of Ogoniness trumping subgroup identities seems to echo the creation of Umuaro from five villages in the face of external threat.

Yet Saro-Wiwa framed the Ogoni identity grounding MOSOP not as an extravagant aberration but as the "genius" of a people that already existed and to which they must return (*Month* 130). As a result, he downplayed the kind of negotiation and sense of agency Achebe represents as necessary to *forge* an effective collective identity characterized by unity in diversity. Without such work, Achebe suggests, conflict and competition will undermine resistance. Some have argued that this was precisely the long-term outcome of Saro-Wiwa's narrative of Ogoniness. Although that narrative may have been a useful tool for mobilization, Michael Watts argues, it contributed to the unraveling of MOSOP into "fragments of class, clan, generation and gender" and to "conflict and struggle among and between ethnic minorities" after Saro-Wiwa's death (Watts, "Violent" 289–90).

In addition, Saro-Wiwa's pastoral narrative of the eco-Indigene elides the question of whether particular environmental attitudes and practices need to be fostered among the Ogoni and within MOSOP. He offers a picture of an Ogoni consciousness aligned with the spirit of local nature in order to indicate that MOSOP and the Ogoni will necessarily protect the ecosystem of the Niger Delta. As I have suggested, this narrative suppresses such issues as the tension between MOSOP's call for greater access to oil royalties and mining rents, which risks perpetuating the attitudes enabling ecological destruction, and their goal of protecting the delta's ecosystem.

In its suggestion of a colonial fall from a rural culture deriving its values and virtues from connection with the natural world, *Arrow of God* also could be described as pastoral. However, it pushes the boundaries of the genre in ways Saro-Wiwa does not. Most obviously, Achebe represents Igbo culture and local nature as profoundly historical and as having been shaped by a mutually mediating relationship. In the novel, to be in touch with nature is to historicize it and to recognize the ways it refuses representational closure. Meanwhile, colonial culture's alienation is defined precisely by its dehistoricizing romance of nature, which encourages a sense of mastery and transcendent knowledge and discourages inquiry, dialogue, and creativity.

Hope in *Arrow of God* is not to be found in the notion of return to an ideal past rooted in a perfect accord with nature or even in the promise of a culture and identity purified of colonial modernity's effects. Achebe's profoundly historical worldview undermines the assumptions underpinning such possibilities. However, the novel does encourage a return to values enabling the formation of strong, diverse socioecological communities. These values include humility, respect for otherness, skepticism regarding knowledge and representation, and a profound sense of responsibility to and dependence on all the parts of a biotic community.

Such values are, of course, also central to Aldo Leopold's land ethic. According to Leopold, ethics depend on the idea that "the individual is a member of a community of interdependent parts" enabling survival, requiring cooperation, and limiting "freedom of action" (202–3). He wants to extend ethics to the entire biotic community and to challenge a "strictly economic" understanding of the relationship between humans and land "entailing privileges but not obligations" (203). His land ethic "changes the role of *Homo sapiens* from conqueror of the land-community to plain member and citizen of it," a transformation that requires relinquishing the sense that one knows "just what makes the community clock tick, and just what and who is valuable, and what and who is worthless" (204).

Arrow of God encourages the kind of ethic touted by Leopold and, in this sense, can legitimately be termed "ecological." At the same time, Achebe remains more attuned than Leopold and many ecologists to the mediation by political relationships of biotic communities and the conceptualizing of them.

Revising Resistance: Tanure Ojaide and Ogaga Ifowodo

With conditions only growing worse since Saro-Wiwa's death, there remains a "gigantic reservoir of anger" in the Niger Delta (Watts, "Sweet" 39). Yet this anger has not been effectively harnessed. Despite its successes in blocking oil production, the Movement for the Emancipation of the Niger Delta (MEND) has contributed to the rise of anarchic violence and failed to articulate a coherent message regarding its goals; even the organization's spokesperson proclaimed that its members were not "revolutionaries . . . just extremely bitter men" (Junger). Meanwhile, state violence and both intra- and interethnic conflict have increased exponentially, and the ecological devastation in the delta remains mostly invisible to an international community focused more on oil security than on human rights or environmental injustice. In their efforts to address growing despair and frustration, the delta poets Tanure Ojaide and Ogaga Ifowodo have drawn on and transformed Saro-Wiwa's narratives of resistance. Like him, they foreground the injustice of neoliberal globalization and stress the possibility of finding inspiration among the peoples of the delta; at the same time, they point to the limitations of ethnic identity, offer an open-ended conceptualization of resistance, and refuse the role of authoritative sage in ways he did not. In the process, they revitalize Saro-Wiwa's project of imagining an alternative trajectory of development in the Niger Delta.

While Ojaide's *Delta Blues and Home Songs* (1998) and *Tale of the Harmattan* (2007) share a similar political and aesthetic commitment, the two collections have some important differences that can be at-tributed to the changes in the decade following the executions of the Ogoni Nine. In the earlier collection, the poet's "blues" result in large part from those executions and what they reflect about Nigeria's condition; at the same time, hope lies in the Ogoni martyrs' message and example and in the promise of a future free of military rule. Published well after Abacha's death and the return to democratic rule, and after almost a decade of growing armed conflict and continued socioecological catastrophe, *Tale of the Harmattan* also finds the poet struggling with despair. It too is part of a Nigerian poetics of "lamentation" expressing "aestheticized rage in the form of sad songs" (Egya, "Aesthetic" 102).[7] However, the reasons for his despair have shifted or are differently inflected as a result of the increasing violence in the delta and the lack of improvement after Nigeria's shift to democratic rule in 2000; in turn, the grounding for hope he offers entails both inspiration by *and* a clear departure from Saro-Wiwa.

Delta Blues and Home Songs laments the degraded condition of Nigeria as a whole and its implications for the Niger Delta. The problem is not simply Abacha, "the usurper-chieftan," but also his "legion of praise-singers," the "uniformed caste of half-literate soldiery," and the corrupt elites with a "voracious love of *naira*" who enabled Saro-Wiwa's murder (42–43). The "bewitched land" finds itself under the curse of the petro-state: "There's an incubus on top of the nation, / wears out the body and smothers smiles" (32). This curse is especially evident in "the delta of [his] birth." The flow of oil "from an immeasurable wound" destroys "this share of paradise"; the poet finds himself overcome by the injustice of having rivers and "sacred soil" ruined by "money mongers" and "thieves" turning his "birth right" into "a boon cake" for themselves (21–22). With the death of the Ogoni Nine, without anyone who can "go further" in the fight against "greed and every wrong of power," he finds the "inheritance" of the delta "now crushes [his] body and soul" (23).

Yet the poet finds hope in the possibility that those inspired by Saro-Wiwa's words will eventually overcome Abacha and the legacy of military rule. From the first poem, he finds himself channeling the dead Ogoni activist-writer: "Now that my drum beats itself, / I know that my dead mentor's hand's at work" (10). Under Saro-Wiwa's spell, he sings of the "unbroken park, / teeming with life" of his childhood, when "the forest green was the lingua franca with many dialects" (12) and of Shell's destruction of the life-giving "bond" between humans and their ecological home (12–13). If before he did not challenge the rhetoric of development "for fear of being counted/in the register of mad ones," under Saro-Wiwa's influence he finds himself openly questioning the wisdom and morality of the delta's annihilation: "so many trees beheaded/and streams mortally poisoned/in the name of jobs and wealth!" (13–14). It may be that after his cowardice he finds himself able to commune only with the spirits of the destroyed flora, "neighbors and providers/whose healing hands of leaves/and weeds have been amputated," but the song he sings also represents hope for an ecological home brought back to life (14). Saro-Wiwa's words still have power; the executioners "conspired to cut his lashing tongue/but they failed" (31).

Similarly, all the Ogoni Nine still live as an inspiration. In "Elegy for Nine Warriors," they walk "back erect from the stake" through the power of memory and enable the living, "those in the wake" who "raise grieving songs," to "look up to the promise of unfettered dawn" (25–26). This "promise" includes Abacha's and his cronies' demise; Saro-Wiwa's name "rising along the dark waters of the Delta/will stir the karmic bonfire" that will consume his nemesis's "dominion" (28). Eventually, insists the speaker, the injustices accompanying military rule will come to an end. "Since the first decree" when they took power, soldiers have benefited while the country has degenerated; however, the "desert's not infinite," and "a stretch of green advances from across the horizon" (51). Such moments clearly align *Delta Blues and Home Songs* with what Sule Egya argues is a "subversive discourse" in Nigerian poetry "aimed at dismantling the rhetoric of military messianism in Nigeria" ("Imagining" 356).

At the same time, the speaker repeats Saro-Wiwa's message of nonviolence. It may be that "many must smart from the sting of [his] songs," but he carries "no weapons or fetish" (36). He rejects "payback" and recoils from the "nerve-wracking" cycle of violence seen in places like Liberia; he laments "Doe's end in which Johnson," the new brutal dictator, tortured the old one to death: "Vengeance is sweet, but that's to/the primitive heart we fight so hard against/in the uniformed chief and herd of sharpshooters" (53–54). Struggling not to stoke "hard grudges into an everlasting blaze," *Delta Blues and Home Songs* echoes Yeats's fear in "Easter 1916" of the heart becoming hard in the midst of revolution; however, in this case the revolutionaries are the symbols of nonviolence rather than armed insurrection (54).

Published almost ten years later, *Tale of the Harmattan* finds Ojaide still "blue," as his former hopes have not been realized. The goat songs of the collection—that is, the songs "of anguish and complaint" (62)—focus on imperial exploitation fueled by a deceptive discourse of progress and salvation. Continuing the work of "pentecostal converts" who "burnt down the primeval grove" in a different colonial age, developers "tore down the forest" and "trashed the natural canopies" in the name of the "foreign-accented god" of modernization[8]: "they argued we needed roads to go out, as if we knew/nothing of adventure or did not visit other places" (12). Although "the government assures people of development," the vast majority is left with neither the riches of the past nor the benefits of modernity. There are no "scholarships" or "jobs for the graduates in the oil sector"; meanwhile, "wells litter the family's farmland," and the speaker's children "can't fish or tap rubber as [he] once did" (22). Foreign capital and a national elite reap their profits at the expense of the livelihoods, the health, and the futures of the delta's people: "We know the capital gain from the blessed but besieged land will go down the drain for a caste to maintain its smug smile" (13). Reversing the development narrative, the narrator claims they have entered a "new Stone Age" of primitive capitalist accumulation "with refilled slave ships refurbished as super-tankers/anchored at Escravos and poaching inland as centuries ago" (22).

Ojaide also uses imagery of monstrous transference to represent the geography of uneven development. A vampirelike process of oil extraction and global distribution is imagined as sucking the life from the Niger Delta: "The blackened stream is ancestral blood/tapped away by giant pipes into ships/to rejuvenate foreign cities, invigorate markets" (10). In "Transplants," the eponymous metaphor figures the process by which the United States benefits from petro-driven modernity and still preserves its ecological riches by creating an environmental apocalypse elsewhere (like some huge portrait of Dorian Gray). In the delta, "the forest fell/foul to fires of oil blowouts and poaching raids," and "the creeks [he] fished in without care" are "now clogged" (39). Meanwhile in the United States the kind of natural beauty and biological diversity he associates with the forest landscape of his youth are preserved: "the pristine streams, the multiethnic population/of plants, costumed birds, and graceful game" (39). Seeing the two landscapes together in the eye of the imagination helps highlight their connection: "In a half-century one world disappeared; another persists./Only outside do I now see the landscape of my childhood" (39). In the context of economic globalization, environmental preservation in the affluent lands of the North cannot be separated from ecological catastrophe in the Global South. Through such double vision, Ojaide subtly undoes pastoral tropes of retreat by emphasizing the impossibility of separating nature from monstrously unjust transnational relationships.

New developments since Saro-Wiwa's death have made the speaker especially aware of the difficulty in reversing the catastrophe in the delta. For example, he now recognizes that the underlying problem is not a particular regime, military dictatorship in general, or the internal dynamics of Nigeria, but economic "globalization," which, like "a category-5 hurricane," baffles efforts to anticipate its direction and "leaves litters in an insane trail" (19). As a result, the return to democracy (in 2000) did not bring about the needed changes; it "demolished monuments of dictators . . . but still we bleed" (19). The speaker also now sees how the people of the delta destroy themselves through interethnic conflict. If in the town of Warri, the location of disastrous "communal violence" (Watts, "Violent" 275), the "Federal Government College" was at one time a place where "every group came together as one," now you "can't escape being cornered into ethnic bunkers" (30). The "bad blood" and "dearth of good will" have transformed the town: "Warri has never really been a beauty / but this defaced figure is not my love" (29). Yet there is no pastoral escape; the city is only part of a disastrous situation in the delta as a whole. When "families" flee they find only "death" and "evacuate the town of all tongues / to join separate militias and be damned" (30). Ojaide is now well aware that ethnic identity is part of, rather than a solution to, the problem; deployed by the militias and associated with the fight over compensation, it has become an idea that leaves the peoples of the Niger Delta "damned" to fighting among and creating unlivable conditions for themselves.

Spiraling socioecological violence makes the speaker wonder in "At the Kaiama Bridge" whether the time for resistance is over. With the "flotillas of river spirits" retreating from "waters . . . turned to a poisonous brew," and with the people turned into environmental "refugees" stuck "in diarrhea-infected camps," he considers the possibility that those fighting for the survival of the delta have failed: "we have organised a resistance army, / declared sovereignty over our resources; / but have not pushed back the poachers" (33–34). The title of the poem refers to a town closely associated with a history of resistance. It was the home of Isaac Adaka Boro, an Ijo revolutionary and hero who tried to create a Niger Delta Republic in 1966 (Okonta and Douglas 149). It was also the place where "the now historic Kaiama Declaration" of 1998 was signed (Okonta and Douglas 145). Over five thousand Ijo youth accused the government and oil companies of severe damage to their people's natural environment and health, claimed ownership over their land and natural resources, demanded the withdrawal of military forces, insisted the oil companies stop all "activities in the Ijo area," and stated that the "youths . . . comprising the Ijo nation would take the steps to implement these resolutions" (Okonta and Douglas 146). When the military realized that the Ijo Youth Council intended to follow through with this declaration (nonviolently), they went on a campaign of murder, looting, and rape. Subsequently, an armed resistance group appeared called "the Egbesu Boys" (Ojaide, *Tale* 62). In his poem, Ojaide asks if the tradition associated with Kaiama is over: "Is revolution dead and must the Egbesu Boys / surrender rights of ownership and humanity?" (34).

In the subsequent poems, the speaker catalogs the ecological "fortune" that has been lost—the "rivers" and "lakes," the "forests" and "farms" (35–39); however, he eventually sets aside his "goat songs" and invokes the Ijo "war-god of born fishers and farmers [Egbesu] / to stand steadfast behind" (41) the Egbesu boys who fight for what Ojaide calls elsewhere an "ecology of justice" that will counter the collusion between the Nigerian government and "multinational corporations to extract . . . oil and gas without regard for the environment or wellbeing of local communities (*Contemporary* 66, 77). "The coalition of global powers" has destroyed the rivers, mangled harvests through gas flaring, poisoned the water and air with "insidious chemicals," and bred "an asthmatic and cancer-prone generation," but "boys" will not "sit and be enslaved" or "die without fighting back" (*Tale* 41–42). Resistance, armed with the spirit of (environmental) "justice . . . will always triumph in the prolonged battle," despite all the forces arrayed against it: "Those who bring a running fight to the iguana / will lose their breath and withdraw before long" (42). The proverb offers a promise of success in the long run through fortitude and patience and sums up the wisdom needed to overcome despair in the face of seemingly hopeless circumstances.

Part of this hope is tied to the memory of resistance in the delta, which has taken different but still connected ethnic manifestations. In his invocation of Egbesu, he references his devotion to the Ijo god's Uhobo equivalent (Ivwri), and he links the Egbesu Boys with the qualities he admired in MOSOP:

> For the same reason I sang praises of Ogoni youths,
> I praise you Egbesu Boys in song—you cannot be
>
> shackled from enjoying your own land's blessings;
> you do the honorable duty of brave sons—fight on. (41)

In making the connection, Ojaide both emphasizes his solidarity with Saro-Wiwa and, at the same time, encourages collective resistance across ethnic boundaries and the use of violence in ways Saro-Wiwa never did.

Finally, the speaker finds hope in the delta itself. It empowers "those who know their land from birth" and who "cannot be pushed out by armed invaders": "the Navy cannot penetrate the fingers of the Niger" (42). It also embodies the spirit of endurance. In a "Dialogue," the "Delta" proclaims to the "Niger" its power to survive any onslaught: "What course stronger than this current [of the Niger] / in hundreds of draughts can drown / the vast calabash that is my dear life?" (44)

Unlike Saro-Wiwa or (to a lesser extent) Ojaide, Ogaga Ifowodo does not answer the question of how socioecological disaster, injustice, and violence in the Niger Delta can be countered. His collection *The Oil Lamp* consists of five poems with scenes or situations initially described by a narrator and with staged dialogues between marginalized subaltern voices and the official scripts of military men and politicians. Any hope in these poems lies in the possibility that answers will emerge from the voices of the oppressed and in the image of the humble, attentive author-activist listening to the wretched of the earth from whose memories resistance can be reborn.

Like Ojaide, Ifowodo depicts the usually invisible violence of uneven development using images of a monstrous modernity. In one poem, a fire at Jese begins after old, corroded pipes break and people rush to collect the spilled "kerosene and petrol" (6). In the blaze, the town and its people are incinerated. An official script attributes responsibility to a criminal and misguided resistance: "a dangerous band of youths sworn to sabotage / for redress of perceived wrongs, / had taken to breaching pipelines" (16). Such accusations have been a long-standing PR strategy; the oil companies and government have often claimed that spills are primarily caused by sabotage and theft rather than by rusting and obsolete pipelines (Okonta and Douglas 72, 78). Writing back, Ifowodo represents the disaster as primarily caused by an imperial process of oil extraction and distribution. The monstrousness of this process is evoked by the image of a "halogen-eyed Cyclops" that was used to light the now defunct oil installation in Jese (4). A promise of electricity for the town was not kept; instead, "the electric Cyclops blinked, moved to another well in another place to guard a fresh promise of light" (4). Without electricity or oil, the town was suffering from a "fuel crunch," which drove the scramble when the pipes broke open (3). The spill itself is caused by "old pipes corroded and cracked" taking oil

> from rotting dugouts and thatched huts
> to float ships and fly planes,
>
> to feed factories and chain of ease
> to heat stoves and save the trees
> to light house and street at break of night,
>
> to make fortunes for faceless traders. (5)

In other words, as in *Tale of the Harmattan,* disaster is brought by a vampirelike modernity taking the wealth of the delta, leaving its people without the means to address their basic needs, and enabling "ease" and "fortunes" elsewhere.

In perhaps his most striking poem, "Ogoni," Ifowodo introduces us to "Major Kitemo." The character is based on Major Paul Okuntimo, who led a notorious army unit, the River State Internal Security Task Force, intended to subject the Ogoni to state authority through any means necessary (Okonta and Douglas 128). In the first six months of 1994, he orchestrated divisions among MOSOP; the murder of the four Ogoni elders for which Saro-Wiwa was arrested; and the subsequent "wasting operations" in which his task force "spread terror, rape, torture, and death, and turned thousands into refugees" in Ogoniland (Okonta and Douglas 129–31). Describing Kitemo as "chief pacifier / of the lower Niger's / still primitive tribes," the narrator marks him as a contemporary manifestation of the colonial administrator parodied at the end of Achebe's *Things Fall Apart* (37). This triple-voiced title (an echo of Achebe's parody) invokes and mocks a neocolonial script depicting the Ogoni as in need of pacification because they are too "primitive" to understand the modern Nigerian nation. In describing the success of having "at last laid waste / the prickly land," Kitemo tells the press how he

> showed what a half-breed the people were
> to claim and fight for what they did not own,
> to deny the owners what was theirs
>
> by decrees duly made and in the books;
> showed why extreme measures were called for
> to teach the needed lesson . . . (37)

The rest of the poem undermines Kitemo's rhetoric by representing him as being taught a "lesson" when he tries to convince the Ogoni to give up their claims and as only subduing their voices through lawless violence.

Assuming the impossibility of grassroots opposition grounded in knowledge and reasoned reflection, Kitemo goes to "the women, fishermen, farmers,/ and jobless youths," who have supposedly been manipulated by "the Ken Saro-Wiwas/ of the foreign sponsored MOSOP" and who must be brought "to reason" (38). However, when he asks them, "Do you really believe you own the oil?" they expose *him* as "brainwashed" (38). He assumes the legitimacy of the nation-state and its "decrees and edicts," which have determined that they do not own the land and its wealth and which make them "citizens/ of a nation, known to law/ and safe from plunder" (40). However, his adversaries question Nigeria's legitimacy. An old man contrasts its recent construction, "born in 1914," and origins in colonial exploitation with the Ogoni's long and close relationship with the land: "How long do you think we have been on this land,/ how long the oil, the trees, the creeks and the rivers?" (39). Meanwhile, a young boy questions the grounding of the state in the will or interests of the people over whom it exercises control. He asks, "in whose name, and by whose powers,/ were the laws you cite made?" (41). The challenge posed by both the old man and the young boy are summed up by a woman asking, *"who/ or wetin make up dis Nigeria?"* (40). The question suggests a deep skepticism regarding the connection between the nation and the two bodies that supposedly ground it—the people and the land.

Bridling at the challenge to his authority and unable to come up with an effective response to "child, man or woman," Kitemo moves from trying to "cure them of the lies/ drilled into their bones by rabble-rousers" to the other methods outlined in his "Memo of Peace" (based on an infamous memo written by the real Major Okuntimo to his superior):

> set them against themselves,
> cause feuds with neighbors,
> and if still true to the rote, still starving
>
> the treasury of petro-dollars, launch
> the last plan: Wasting Operations,
> group infiltration, mass deportations. (43)

"Unspooling / threads of suspicion" and strengthening them using "bribes / and slander," he fosters the killing of the four Ogoni elders, which served as a pretext for the execution of Saro-Wiwa and the Ogoni Nine, as well as for the "Wasting Operations" by the security forces, which included murder, rape, torture, a massive refugee crisis, and starvation (47). These "operations" are Kitemo's means of mastering those who would dare challenge him, "the smart-ass quartet: the old man, the schoolboy / and his father, the woman": "They came out among the crowd and fell on their knees: / *Please, please, we will do what you say, anything / you want. But stop the shelling. Please! Please!*" (48). However, if at the beginning and ending of the poem Ifowodo points to a supposed silencing of the Ogoni through state violence, the poem itself still evokes their voices to challenge the official discourse of the state embedded in Kitemo's stock pronouncements. Supposedly marking the inferiority of those he has come to convince, double-voiced phrases like "the hard rote of lawlessness" and "the strength of their delusion" come to apply to him and his kind (39–40).

Allegorically, Kitemo embodies the Nigerian petro-state constituted by empty signifiers of legal authority, and the Ogoni characters are the voice of a dissent struggling to turn that state's discourse of criminality on its head. They point to a potential foundation for a different kind of nation, one grounded in the dialogic performance of the people and in identification with and protection of the land. In one sense, they could be understood as an expression of "the 'organic' idiom" in which MOSOP's struggle was framed and that "developed into a *political ecology* of citizenship for all Nigerians" (Apter 260). Yet Ifowodo does not focus on the articulation of Ogoni autonomy or, more generally, on the specific aspects of MOSOP's political agenda. Instead, he emphasizes the need for the artist-activist to go to the voices of popular dissent to find grounding for hope and the way to a better future.

At the same time, the promise represented by the wisdom and defi-

ance of the "women, fishermen, farmers, / and jobless youths" is severely
circumscribed by the sense of despair stemming from their deaths and
silencing by the violence of Kitemo. This despair pervades the next poem,
"The Pipe Wars," as the voice of a "bureaucrat's well-oiled tongue" jus-
tifies the continued "pacification" of the Ogoni (51). He too echoes the
petro-state's positioning of the delta and its people as inferior and uncivilized and of the Nigerian government as the means to modernize the
nation through distribution of oil revenue:

> so absurd
> their claim for redress, that it should empty
> the coffers and deny the nation's engine
> its lubricant. Rust would follow; there'd be an end
>
> to motion and a nation to call our own. (53)

However, unlike in "Ogoni," the voice of official national discourse is
not challenged. Instead, the bureaucrat describes the turning of anger
inward as communities fight among themselves for the crumbs left by
the oil industry: "a scrapping for used pipes" became "a war for territory /
between Oleh and Olomoro" and serves the state as a further "pretext /
to shoot and burn, and drive more into the bush" (54). Meanwhile, livelihoods and health continue to deteriorate. In the final section of the
poem, the narrator describes the Ogoni as struggling "for a living / where
the land's promise was boundless ease," while "Iron-Dragon— / the
gas-flaring stack whose awful mouth spits fire / without cease" retches
"on every head afflictions and deaths / sucked from the depths of the
earth" (55). The oil industry and the Nigerian petro-state pour waste on
the delta and turn it, as the title of the final poem indicates, into a "cess-
pit" (57).

In this poem, consisting of "a few more songs" begged by "the
many scars that itch and wounds that bleed / far from the eyes of the
world," the narrator mostly offers scenes of despair and injustice (59).
He sings of children needlessly dying because there are no clinics "in a
hundred miles" and "no motorway" to get them to one quickly enough
(60). Meanwhile, "the oil staff estates" are "well-drained and paved and
mosquito-proof" with "carpet-lawns" and "quiet order" (62). One "man
with a grudge in Warri," shamed by the discrepancy between these "estates" and his "shack in the swamp" without electricity, comes to feel
himself worthless, "like a sewer rat," and is willing to separate him-

self from the community—go "solo"—in pursuit of the wealth of the oil industry:

> That's when I said, "Listen, Soloman,
> you're nobody's fool. Be wise! Good job,
> fine house, sleek car and beautiful woman
>
> are for Solo too. That's your money, man!
> And I'm going to live even if I die first" (62)

In a sense, this moment is the nadir of *The Oil Lamp*. (Solo)man represents all those who have given up hope for a different world and are willing to instrumentalize themselves for the oil industry.

However, the narrator refuses to despair and to relinquish his "naked malice" toward those turning the delta into a "cesspool" (63). He denies "Major Kitemo (and his ilk) the last word" by giving it to the Major's "nemesis, the schoolboy,"

> now a sophomore at law, unaware
> what paths to even sharper perception
> the gods, offended by bombs, prepare
>
> for his itching feet as he writes under a tree
> —and I copy—the words of the shaping poet,
> peeling the scabs to regrow the skin. (63)

The narrator does not position himself as leader or sage. Instead, he takes on the more humble role of recorder and articulates the *hope* of new "paths" to "even sharper perception" stemming from acts of creativity and memory. "The shaping poet" envisions the petro-state as a sick body spewing its corruption, "the pus of the land," down the "two valves" of the Niger "East and West of the putrid heart" into "our open veins." However, the people of the delta will not be overcome but find "footholds" and "stay afloat" through the memory of injustice—"the humus of hate and envy." This memory enables them to resist the stale rhetoric of "Kitemo (and his ilk)": "And when they rise to spit on our heads the rinse-water / of their morning mouths, I remember the dew, / the one thousand and one gone, and what will remain true" (63). By holding onto the memory of resistance and of hope, the poet keeps alive forms of representation challenging the deadening official lies and maintains grounding for a different future.

✳✳✳

Faced with growing socioecological violence and despair, Tanure Ojaide and Ogaga Ifowodo overtly align themselves with Ken Saro-Wiwa and, at the same time, deviate from a number of his conclusions. In this sense, they are engaged in a process of what Henry Louis Gates calls "signifyin(g)." According to Gates, signifyin(g) "functions as a metaphor for formal revision, or intertextuality" (xxi); it is repetition with a difference. If Ojaide's and Ifowodo's revisions foreground homage and continuity (rather than disjuncture), they still, like Achebe's model of cultural production, entail significant (perhaps even extravagant) aberration.

conclusion

THE QUESTION OF HOW TO CONCEIVE PLACE IN RELATION
to the politics of scale represents a conceptual challenge for environmental justice narratives. Such narratives often encourage skepticism regarding "transcendent and universal politics," which potentially marginalize situated perspectives and hamper the formulation of place-based identities crucial for local mobilization (Harvey, *Justice* 400). At the same time, the history of environmental justice activism suggests a need to link local movements with forms of resistance and resistant identity operating at larger scales. In the context of political ecology, David Harvey sums up the challenge in terms of a geographical dialectic in which narratives produced at "local scales" are not "in some way subservient to the larger story," nor is "locality (place) where the unique 'truth of being' resides." The problem, according to Harvey, is that "we have yet to come up with satisfactory or agreed-upon ways to do this" (*Cosmopolitanism* 229).

In some ways, Aletta Biersack's notion of "a place-based approach," drawing on the work of Doreen Massey, goes some way toward addressing the challenge (17). It conceptualizes place as *the grounded site of local-global articulation and interaction*" (Biersack 16). According to Massey, place needs to be formulated in terms of "the particularity of linkage to that 'outside' which is therefore itself part of what constitutes place" (67). Its specificity derives from the ways *a* place is "located differentially in [a] global network," "from the fact that each place is the focus of a distinct mixture of wider and more local social relations," and from the way "all these relations interact with and take a further elements of specificity from the accumulated history of a place" (qtd. in Biersack 16). This approach to place, claims Biersack, helps political ecology avoid a reliance on "dependency theory and world system theory," which subordinate the local "to a global system of power relations" and necessitate that it "be understood entirely with respect to that subjection" (9). In contrast, a "place-based approach" enables a "rotation from a vertical and binaristic to a horizontal and dialectical perspective on local-global relations" (17) and is especially important in terms of acknowledging "the importance of grassroots agency" (19).

This kind of approach does not exactly bring closure to the question of a "satisfactory" way to conceive the relationship between place and what is positioned as outside. For example, if it addresses the ways that "transnational space and place *co-arise*" to some degree, it still starts with and, as a result, potentially privileges place (18). The focus on a "horizontal" axis also risks downplaying the often incredible vertical power of processes and uneven relationships working at larger scales.

Nonetheless, the postcolonial regional particularist angle taken in this book is in many ways aligned with Biersack's place-based approach. Postcolonialism has been highly attentive to the ways imperial discursive and material power generates difference and to how such difference matters for understanding various kinds of unequal political and economic relationships and the possibilities for their transformation. As a result, postcolonialism must always face the challenge of balancing attention to particular differences with attention to the relational construction of place, culture, and identity through asymmetrical power.

The regional particularist approach described in the first chapter encourages a consideration of how Africa might be thought about as different from other parts of the world in terms of the way it has been shaped by imperialism and globalization. According to James Ferguson, the history of the interrelationships among (imperial) power, economic processes, and representations of Africa has made "a category that (like all categories) is historically and socially constructed" also one "that is 'real,' that is imposed with force, that has a mandatory quality" (*Global* 5). Africa's association with savage wildness has been a significant factor in the continent's transformation into what he terms a "place-in-the-world." This association has facilitated the undermining of Africans' agency and humanity, patterns of unjust extractive and ecological enclaves, and socioecological transformation on a massive scale. The African environmental literary traditions examined in this book draw attention to the relationship between these conditions and stereotypical naturalizing representations of the continent.

However, postcolonial regional particularism will not only be attentive to shared characteristics of a place in the world but also pay close attention to differences within such a place. Africa may be a place of the sort described by Ferguson, but it is obviously profoundly heterogeneous and constituted by other (kinds of) places. The African texts reflect this heterogeneity, as do the different parts of the book. Chapter 2 focuses on the subcontinental region of East Africa, chapter 3 on a nation (South Africa), and chapter 4 on something like a bioregion (the lower Niger).

One way the texts differ is their very conceptualization of place and its relationships with history, other places, and geographic scale. Attending to these differences and their significance offers a fruitful way of approaching the challenge of conceptualizing place and the politics of scale for environmental justice struggle. Some of the authors embrace the possibility of conserving or returning to transhistorical and centered place-based identities embodied in an ideal past and/or in traditional rural landscapes. Others suggest that such (pastoral) representations run the risk of eliding complex divisions and connections shaped by history and, concomitantly, of hampering the bridging work needed to effectively represent common interests and forge resistance. These authors can often be aligned with an open-ended, nonteleological version of geographical dialectics. For example, Zakes Mda embraces an exploratory approach to imagining relationships across scale and for skepticism toward representational closure.

Like *Heart of Redness*, Paton's *Cry, the Beloved Country* and Head's *When Rain Clouds Gather* point to cases of environmental injustice and the need to connect conservation with political relationships. Yet they also rely on stable, autonomous conceptions of place in ways Mda does not. The local "solution" *Heart of Redness* depicts may not adequately address the threats posed by economic and political relationships operating at larger scales, but Mda brings attention to the limitations of this solution and suggests that balancing local perspectives with efforts to think across scale is an ongoing process *in* environmental justice struggles that can never offer an abstracted, final word on praxis.

Putting Ken Saro-Wiwa's activist narratives in intertextual dialogue with earlier and later writing from the lower Niger also highlights possible dangers in transhistorical, autonomous conceptions of place and collective identity for environmental justice struggle. Saro-Wiwa was extremely effective at bringing attention to political and economic relationships working at different scales and at mobilizing resistance across scale. Although *Arrow of God* can be connected to his writing through its emphasis on the environmental injustices initiated by imperial development, Saro-Wiwa focuses much more explicitly than Achebe on uneven political and economic scalar relationships. However, *Arrow of God* offers a nuanced way to question Saro-Wiwa's representations of transhistorical platial identity through its emphasis on the need for creative transformation (rather than return). Current conditions in the Niger Delta also suggest a need to revise, as well as reiterate, Saro-Wiwa's narratives. Bringing together his ability to think across scale with Achebe's notion of extravagant aberration might be one way to achieve such revision.

However, in many circumstances the type of transhistorical identity embraced by Saro-Wiwa may be not just useful but also necessary for the forging of effective resistance. Such conundrums cannot be addressed in the abstract; they must be worked through in the context of particular environmental justice struggles and different moments in these struggles. In this regard, the echoes and transformations of Saro-Wiwa's narratives in Tanure Ojaide's and Ogaga Ifowodo's poetry take on particular significance. Confronted with increasing socioecological violence and despair, Ojaide and Ifowodo follow Saro-Wiwa's lead by foregrounding the need to represent relationships across scale to understand environmental injustice in the Niger Delta; however, they also bring into question the benefits of an emphasis on ethnicity and conclusive representations of resistant identity and leadership. In the process, they both address the limitations of Saro-Wiwa's narratives highlighted by developments since his death *and* emphasize the continuing relevance of those narratives.

Questions of place and communal identity also point to conflicting perspectives on the category of nature. The traditions of environmental writing I discuss challenge hegemonic assumptions regarding development, conservation, and nature, as well as the separation of ecological projects from sociopolitical relationships. At the same time, they suggest that survival and progress for those on the losing end of global imperial development require a move away from the instrumental approach to ecological relationships and processes and the lack of humility often entailed by such development. However, the texts also differ significantly in terms of underlying assumptions regarding nature and human relationships with it. In contrast with many political ecologists' rejection of naturalizing narratives, authors such as Saro-Wiwa, Maathai, Okot p'Bitek, Ngũgĩ wa Thiong'o, and Camara Laye allude to or depict precolonial societies that were in harmony with a local nature enabling a unity of identity and a conflict-free existence. They offer hope of rejuvenation through a return to an indigenous form of dwelling based on an accord with a historically transcendent natural spirit of place. Other authors historicize nature and question the existence of natural essences that can ground identity and to which communities need to reconnect. In *Arrow of God,* one danger of colonial discourse is precisely its suppression of the heterogeneity and sociohistorical aspects of nature and place. Achebe may suggest that colonial modernity brought alienation, but this alienation is understood as stemming from an imperial romance projecting a static and singular nature separated from history and culture. Texts like *Arrow of God* can be aligned with political ecology's often skeptical approach to romantic representations of community and nature generated in the context of anticolonial and environmental justice struggles.

Bringing different deployments of anticolonial pastoral tropes into dialogue draws attention to the significant questions raised by conflicting representations of nature and naturalized identity for such struggles. Reading Maathai's autobiography and essays, Okot's poetry, and Ngũgĩ's *A Grain of Wheat* contrapuntally points to an African literary tradition drawing on such tropes and aligned with environmental justice; however, it also highlights potential drawbacks of these tropes. For example, they can result in oversimplified solutions to current environmental problems through the elision of structural relationships and in notions of authenticity legitimating inequalities and injustice based on gender, ethnicity, caste, and so on. These dangers are highlighted in Nuruddin Farah's *Secrets.* Although Farah's novel has a strong environmental focus, it suggests that claims to speak in the name of nature (to know its "secrets") all too easily help (re)produce unjust power relations and exploitation. Yet, if *Secrets* brings attention to dangers in pastoral representation, reading it in relation to Maathai's writing and the history of the Green Belt Movement also foregrounds potential limitations in Farah's antipastoral vision and his politics of difference. Maathai used her historical narrative of indigenous environmental practice and identity in order to transform consciousness, establish socioecologically progressive praxis, and mobilize political resistance; the history of the Green Belt Movement suggests benefits in such a strategy.

In a different context, Steve Biko advocated for the development of "Black Consciousness" based on the writing of "a positive history" that would challenge negative colonial images of "the dark continent," undermine "close identification with the white society" of South Africa, and, more generally, transform "standards and outlook" (29–30). Such a history, he argued, would contribute "to meaningful and directional opposition" and to "a singularity of purpose" in the struggle against apartheid (31). The history of postapartheid South Africa might point to potential problems with this position and to the need for different formulations of consciousness, but that does not negate the benefits of Biko's influence at the time he wrote. Evaluating strategies used to formulate collective consciousness, including uses of pastoral representation, needs to be contextual. We must ask, for example, to what degree representations of difference and/or unity and of nature and relationships with it serve "socially emancipatory or disabling ends" in particular situations (Chrisman 198).

In the context of South African history, the need to situate the means (and ends) of resistance is highlighted through a contrapuntal reading of Nadine Gordimer's *The Conservationist* and *Get a Life*. The two novels are clearly linked by a critical vision of colonial conservation associated with environmental justice struggle and political ecology; however, the possibility that ecological science and environmental activism might contribute to the fight for a more just future is foreclosed in the earlier novel and opened up in the later one. The similarities and differences between them in this regard are best understood in terms of continuities and changes in South African environmentalism since the 1970s, but they do not necessarily need to be framed using narratives of enlightenment (or lack of it). Instead, the relationship between the two novels can draw attention to the heterogeneity of environmental justice struggle and representation as it unfolds in different places and moments.

In general, a text's environmentalist or visionary credentials cannot necessarily be measured by how close it comes to (or how far it deviates from) a single (ur)narrative of effective resistance to imperial processes creating unjust socioecological geographies. We can, instead, gain an open, varied sense of the struggle for environmental justice in Africa by putting texts in dialogue and by considering how they contribute to productive but always limited visions of such struggle. This model of reading holds in abeyance forms of closure as it draws on intertextual dialogue as a means to bring to consciousness the socioecological unconscious of narratives offered by political ecology or developed in the context of environmental justice struggles. At the same time, the contrapuntal readings throughout this book draw attention to connections among various African literary texts based on a challenge to malleable (but always related) unjust forms of imperial development and extraction that help position Africa as a certain kind of place in the world.

notes

Introduction

1. Adams and McShane; Adams, "Nature"; Anderson and Grove; Bienart, "Soil," "African," *Rise*; Brockington; Carruthers, "Nationhood," *Kruger;* Fairhead and Leach; Ferguson, *Anti-Politics*; Hulme and Murphree; Leach and Mearns; Neumann, *Imposing, Making.*

2. Or, as William Beinart claims, "measuring change in terms of movement away from a pristine environment, and calling all change degradation, is of limited value. Human survival necessitates environmental disturbance, nor is nature in itself static" (*Rise* 390).

Greg Garrard points out that many ecocritics also subscribe, mistakenly, to notions of ideal ecological equilibrium; he claims that even as contemporary ecology brings a static "balance of nature" model into question, "the association between biological diversity, ecosystem stability and an ideal, mature state of nature is an article of faith for most ecocritics and philosophers" (57).

3. The qualifier *global* marks an "expanded and diversified" conception of environmental justice not subsumed by a racial or nationally bound focus (Walker 2).

1   The Nature of Africa

1. O'Brien, "'Back'"; Cilano and DeLoughrey; Vital, "Toward."

2. As DeLoughrey and Handley point out, upholding "a sense of alterity while still engaging a global imaginary" requires "engaging local and often inassimilable aspects of culture and history" (28).

3. For ecocritical anthologies focused on Africa, see Okuyade; Caminero-Santangelo and Myers; and Wylie.

4. For overviews of American and British ecocriticism see Buell, *Future;* Garrard; and Heise, "Hitchhiker's."

5. Cilano and DeLoughrey; DeLoughrey, Gosson, and Handley; DeLoughrey and Handley; Huggan; Huggan and Tiffin, "Green," *Postcolonial*; Nixon, "Environmentalism," *Slow;* O'Brien, "Articulating"; Tiffin; Vital, "Situating," "Toward"; Vital and Erney; L. Wright.

6. Cilano and DeLoughrey; Huggan and Tiffin, *Postcolonial.*

7. Nancy Stepan notes, "the transfer to natural history, geography, and anthropology of the political terminology of the eighteenth century—'kingdom,' 'nation,' 'province,' and 'colonist'—indicates just how closely the notion of a distinctive tropical nature was tied to political empire" (17).

8. In an early article on ecocriticism and South African literary studies, Julia Martin noted the tension between "a definition of environmental priorities that

was perfectly in keeping with the . . . colonial project" and the concerns of the "the majority of South Africans," who would see such priorities as "irrelevant, and even inimical, to the struggle for social and political justice" (3, 1). In this context, she found striking "a rather uncritical focus on 'nature writing' " in British and American ecocriticism and pondered if "opening the canon to other voices" might "subvert the genre's fundamentals": "Is the nature of Third World environments likely to produce the texts of wilderness, forests and the great outdoors with which we are familiar? I think of the difficulties of teaching Wordsworth to students from the townships" (4).

9. William Finnegan outlines the many limitations of *The Shadow in the Sun* in an impressive book review.

10. Ngũgĩ wa Thiong'o has called *Out of Africa* "one of the most dangerous books ever written about" the continent and particularly abhors Blixen's racist equation between Africans and animals (*Moving* 133).

11. As Simon Lewis puts it, "Karen Blixen is . . . able to present her lifestyle on her farm in Africa as the acme of a kind of natural, or at least extrasocial, civilization," rather than as the product of (brutal) colonial processes (113). For the history of colonial dispossession and exploitation in Kenya, see Ward; Wolff; and Kanogo.

12. Of British heritage, Leakey was appointed head of Kenya's Wildlife Conservation and Management Department in 1989 by Daniel Arap Moi. He "lobbied hard to have the elephant declared an endangered species" and was able to garner significant international financial support for Kenyan wildlife conservation (Bonner 132–33). Raymond Bonner surmises that part of his enormous reputation as a successful conservationist resulted from "the Western desire for a white hero" (133).

13. Adams and McShane; Brockington; Carruthers, "Nationhood," *Kruger;* Hulme and Murphree; Neumann, *Imposing, Making.*

14. Anderson and Grove; Brockington; Carruthers, "Nationhood," *Kruger;* Neumann, *Imposing;* Shetler.

15. Adams and McShane; Adams, "Nature"; Carruthers, "Nationhood," *Kruger;* Hulme and Murphree; Neumann, *Imposing, Making.*

16. See Ribot and Oyono; Chabal.

17. See Derman, Odgaard, and Sjaastad; Hyden.

18. See also Bassett and Crummey.

19. Until recently, "within the English-speaking academy, the experiences of the USA, Australia and the UK have dominated discussion and theoretical development, albeit with important differences being identified between these countries" (Williams and Mawdsley 660).

20. Walker claims that "as environmental justice globalizes, its initial meaning derived from the U.S. context is not simply reproduced, although neither is it entirely abandoned" (33).

21. Nixon's study of the nexus between temporality and environmental conflict is crucial for addressing what Nancy Peluso and Michael Watts consider the "undertheorized" aspect of scholarship exploring "the ways different forms of violence systematically figure in environmental struggles" (6).

2   The Nature of African Environmentalism

1. Spivak, "Can"; Loomba.
2. Beinart, "Soil"; Gadgil and Guha; Ferguson, *Anti-Politics*; Leach and Mearns.
3. Plumwood, *Feminism;* Soper; Bullis.
4. Alden and Tremaine; Sugnet; Myers.
5. Bullis; Plumwood, *Environmental;* Warren.

3   The Nature of Justice

1. McDonald, "What"; Debbane and Keil; Kalan and Peek.
2. Adams, *Green,* "Nature"; Beinart and Hughes; Grove; Robin.
3. See also Beinart, "African"; Leach and Mearns.
4. Hallowes; Hallowes and Butler.
5. See, e.g., Graham; D. Head; Huggan and Tiffin; Vital, "Situating," "Toward," "'Another'"; L. Wright.
6. Beinart, *Rise;* Beinart and Coates; Beinart and Hughes.
7. According to Doug Aberley, a "bioregional world-view" includes the position that "the root cause of [current social and ecological crises] is the inability of the nation-state and industrial capitalism . . . to measure progress in terms other than those related to monetary wealth, economic efficiency or centralized power" (36).
8. Nixon, *Slow;* Buell, *Future.*
9. L. Brown; Clayton; Sample.
10. Coreen Brown claims, "There is sufficient evidence in Head's written accounts, both published and private, to show that Head grew to love the Botswanan landscape, and throughout her writing her evocation of the natural world is an affirmation of her belief in the restitutive quality of the natural as an antithesis to the social and the material" (53).
11. Rob Nixon argues that for Head "land and soil offered the prospect of compensatory affiliations, and came to be inflected as alternative sources of lineage and belonging" in "her quest to replace the hollow organisms of family, race, and nation that had betrayed her" (*Homelands* 123).
12. Guha 107–8; Gadgil and Guha 89, 190; Nixon, *Slow* 140.
13. For an extended endorsement of Head's vision of agriculture in Botswana, see Maureen Fielding's essay "Agriculture and Healing."
14. See the previous section for an overview of gendered nationalist discourse in colonial and postindependence contexts.
15. A full analysis of the complicated intertextual relationship between *The Heart of Redness* and *The Dead Will Arise* is beyond the scope of this chapter. Yet it does seem important to address, however briefly, the recent charge of "plagiarism" against Mda by Andrew Offenburger. Based on a fairly exhaustive catalog of Mda's borrowings from Peires, Offenburger claims that the novel is "a derivative work masquerading plagiarism as intertextuality" (164) and accuses literary critics of abetting the crime. However, his argument that *"The Heart of Redness was formed* by Jeff Peires's *The Dead Will Arise"* (165) only considers the part of the novel placed in the past and, as a result, does not address the relationship between

the historical material and the novel's narrative of the present. Offenburger also ignores the fictional aspects of Mda's historical narrative (the story of the brothers Twin and Twin-Twin), as well as the oral sources on which Mda draws. Finally, Offenburger fails to define key terms such as *plagiarism* and *intertextuality* and to address their rich theoretical history. (For Mda's rebuttal to Offenburger, see "A Response.")

In contrast, Jennifer Wenzel develops an insightful reading of Mda's appropriations of *The Dead Will Arise*. While acknowledging the extent of these appropriations and some of their troubling implications, she ultimately argues that "what makes *The Heart of Redness* more than a novelization of Peires's history is its engagement with questions of time" (179). This reading is part of Wenzel's larger critical project of exploring how "cultural afterlives of the cattle killing"—i.e., how it has been represented in subsequent narratives—trouble notions of linear time and progress, as well as the meanings and heterogeneity of both past and present (27).

It should also be noted that *The Dead Will Arise* has been sharply criticized by historians such as Jeff Guy and Helen Bradford, especially in terms of Peires's dismissal of a gendered analysis. (The evidence is overwhelming that gender and gender oppression were central causal factors.)

16. In this sense, *The Heart of Redness* echoes Lawrence Buell's assertion that "place is not just a noun but also a verb, a verb of action" (*Writing* 67).

17. See also Jacobs; Koyana; Woodward.

18. For the history of colonial attempts to do away with collective land tenure following the cattle killing, and specifically the power of the chiefs to allocate land, see Peires 290–96.

19. Laura Wright argues, "Quekezwa . . . offers a highly flawed and shortsighted solution to the problem of white intrusion; because the landscape has been altered as a result of European influence, removing all invasive species would not only be impossible, but also such action can only operate at a metaphoric level, functioning as a symbolic displacement of a firmly entrenched capitalist development system" (51).

20. For an insightful reading of ways that Mda treats the nature/culture binary, see Klopper. Using "anthropological research on the role of diviner-prophet in Xhosa society," Klopper argues that *The Heart of Redness* interrogates the relationship between nature and culture through the notion of the prophetic (92).

21. See, e.g., Doreen Massey's notion of a "progressive concept of place," which sees "the uniqueness of place" as stemming precisely from being "a particular, unique, point" at the intersection of "wider and local social relations" that themselves "take a further element of specificity" from history (322–33).

22. In *Sense of Place and Sense of Planet*, Ursula Heise asserts the need for what she terms "eco-cosmopolitanism," which entails a shift from the primary focus on place in many forms of American environmentalism and ecocriticism to ways of representing environmental crisis in global terms. While not entirely dismissing the usefulness of the category of place, she seeks forms of environmental imagination and advocacy based "no longer primarily on ties to local places but on ties to territories and systems that are understood to encompass the planet as a whole"

(10). In *Bringing the Biosphere Home,* Thomashow also engages in the project to develop means of grasping global environmental threat. However, he claims that a global environmental sensibility must stem from a sense of place; Heise rejects this position: "The challenge for environmentalist thinking . . . is to shift the core of its cultural imagination from a sense of place to a less territorial and more systematic sense of planet" (56). Even though *The Heart of Redness* emphasizes the ways that the local must be understood in term of specific larger national and global relationships, it too does not quite line up with Heise's notion of eco-cosmopolitanism since, as my reading suggests, the category of place is still privileged.

23. See also Wagner.

24. As Barnard argues, "While the pastoral idea of the local solution is certainly expressed in the novel, the overarching artistic and ethical purpose of the text—one in which the reader is invited to participate—is to construct a new whole, by discovering the relationship between things" (78).

25. In actuality, 26 percent of the profits from the Pondoland mining project are slated for a Black Economic Empowerment (BEE) company. The result is support from local ANC politicians, but not from most of the local inhabitants (Dellier and Guyot 94).

4   The Nature of Violence

1. Douglas et al.; Rowell, Marriott, and Stockman 172–207; Watts, "Violence."

2. For example, after Saro-Wiwa's execution and facing calls for a boycott, Shell ran a campaign entitled "Profits and Principles—Does There Have to Be a Choice." Projecting concern with human rights and environmental issues, the campaign was intended to co-opt the company's critics (at least outside the delta). The two main themes were openness and dialogue; members of "special publics"—the media, NGOs, financial analysts, academics, and government officials—were invited to give their input in multiple venues and told their views would be taken into consideration (Rowell, Marriott, and Stockman 120–24). Similar campaigns followed. The claims made regarding corporate ethics, environmental concern, and community development have been shown by study after study to be hollow, but by 2004 "Shell had been reborn. The perception . . . on the part of the special publics and many of the staff was that Shell was now fundamentally different from the company of nine years earlier" (Rowell, Marriott, and Stockman 129).

3. Even as Shell took "out full-page advertisements in the Nigerian dailies to suggest a descent into terrorism," they were supplying "weapons, through a variety of sophisticated fronts, to security operatives and mercenaries (including local youth)" that they retained (Douglas et al. 246–47). The oil industry watchdog group Platform recently published a report documenting Shell's contribution to "armed conflict in Nigeria by paying hundreds of thousands of dollars to feuding militant groups," many of them known criminal gangs. The company would give money to whichever group could prove they had the means to access its pipelines, wells, and flow stations and, as a result, encouraged savage conflicts over protection money (D. Smith).

4. Apter; Osaghae; Quayson; Raji; Watts, "Violence."

5. Describing a situation eerily reminiscent of this episode, but with an inter- rather than intraethnic twist, Saro-Wiwa claims that "an Assistant District Offi- cer" would "parcel out Ogoni land to their Igbo neighbors" and suggests that this injustice resulted from the rejection of colonial culture by the Ogoni (*Genocide* 16).

6. Michael Lundbland claims that although *Arrow of God* exposes how claims to speak for nature can become malignant fictions, Achebe still deals with animal sacrifice in a way that makes "different narratives that are based upon domination seem natural" and undermines the potential for "ethical interactions with the en- vironment" through "a logic of opportunistic adaptation, which threatens to sacri- fice an appreciation for nature on the altar of development" (9). Equating Achebe's point of view with Ezeulu's and other characters' justifications for their actions, Lundblad pays inadequate attention to the various ways that *Arrow of God* brings into question multiple discourses of development and sacrifice.

7. Many of the chapters from Okuyade's *Eco-Critical Literature* focus on Ojaide's writing, especially *Delta Blues and Home Songs, The Tale of the Harmat- tan, Daydream of Ants and Other Poems,* and his novel *The Activist.* Other literary texts from and about the Niger Delta addressed by the anthology include Kaine Agary's novel *Yellow-Yellow,* Isadore Okpweho's novel *The Tides,* and poetry by Hope Eghagha and Gilbert Ogbowei.

8. Like Maathai, Ojaide has claimed that environmental degradation in Africa results from the undermining of precolonial animistic belief, which "held aspects of nature sacred" and in which groves of trees, in particular, were "an integral as- pect of . . . spirituality": "with the coming of Christianity and Islam to Africa, the natural world became a servant of man rather than a partner because of an aloof God, leaving man to control and exploit nature" ("Foreword" vi).

works cited

Aberley, Doug. "Interpreting Bioregionalism: A Story for Many Voices." *Bioregionalism*. Ed. Michael Vincent McGinnis. New York: Routledge, 1999. 13–42.

Achebe, Chinua. *Arrow of God*. New York: Anchor, 1964.

———. *The Education of a British-Protected Child*. New York: Knopf, 2009.

———. *Hopes and Impediments: Selected Essays*. New York: Anchor, 1988.

———. *Things Fall Apart*. New York: Anchor, 1959.

Adams, William. *Green Development: Environment and Sustainability in the Third World*. New York: Routledge, 2001.

———. "Nature and the Colonial Mind." *Decolonizing Nature: Strategies for Conservation in a Post-Colonial Era*. Ed. William Adams and Martin Mulligan. London: Earthscan, 2003. 16–50.

Adams, William, and David Hulme. "Conservation and Community: Changing Narratives, Policies and Practices in African Conservation." *African Wildlife and Livelihoods: The Promise and Performance of Community Conservation*. Ed. David Hulme and Marshall Murphree. Oxford: James Currey, 2001. 9–23.

Adams, William, and Thomas McShane. *The Myth of Wild Africa: Conservation without Illusion*. New York: Norton, 1992.

Adamson, Joni, Mei Mei Evans, and Rachael Stein, eds. *The Environmental Justice Reader: Politics, Poetics, and Pedagogy*. Tuscon: U of Arizona P, 2002.

Alden, Patricia, and Louis Tremaine. "How Can We Talk of Democracy? An Interview with Nuruddin Farah." *Emerging Perspectives on Nuruddin Farah*. Ed. Derek Wright. Trenton, NJ: Africa World Press, 2002. 25–46.

Anderson, David, and Richard Grove, eds. *Conservation in Africa: People, Policies, and Practice*. Cambridge: Cambridge UP, 1987.

Apter, Andrew. *The Pan-African Nation: Oil and the Spectacle of Culture in Nigeria*. Chicago: U Chicago P, 2005.

Artestis, Philip. "Furor on Memo at World Bank." *New York Times*. February 7, 1992.

Barnard, Rita. *Apartheid and Beyond*. Oxford: Oxford UP, 2007.

Barnett, Clive, and Dianne Scott. "Spaces of Opposition: Activism and Deliberation in Post-Apartheid Environmental Politics." *Environment and Planning A* 39.1 (2007): 2612–31.

Bassett, Thomas, and Donald Crummey. "Contested Images, Contested Realities: Environment and Society in Africa's Savannas." *African Savannas: Global Narratives and Local Knowledge of Environmental Change*. Ed. Thomas Bassett and Donald Crummey. Oxford: James Currey, 2003. 1–30.

Bazin, Nancy, and Marilyn Seymour. *Conversations with Nadine Gordimer*. Jackson: UP of Mississippi, 1990.

Beinart, William. "African History and Environmental History." *African Affairs* 99 (2000): 269–302.

———. *The Rise of Conservation in South Africa: Settlers, Livestock and the Environment 1770–1950*. Oxford: Oxford UP, 2003.

———. "Soil Erosion, Conservation and Ideas About Development: a Southern African Exploration, 1900–60." *Journal of Southern African Studies* 11.1 (1984): 52–83.

Beinart, William, and Peter Coates. *Environment and History: The Taming of Nature in the USA and South Africa*. New York: Routledge, 1995.

Beinart, William, and Lotte Hughes. *Environment and Empire*. Oxford: Oxford UP, 2007.

Bhabha, Homi. *The Location of Culture*. New York: Routledge, 1994.

Biersack, Aletta. "Reimaging Political Ecology: Culture/Power/History/Nature." *Reimaging Political Ecology*. Ed. Aletta Biersack and James Greenberg. Durham: Duke UP, 2006. 3–40.

Biko, Steve. *I Write What I Like*. London: Bowerdean, 1978.

Blixen, Karen (pseud. Isak Dinesen). *Out of Africa and Shadows on the Grass*. New York: Vintage, 1937.

Boehmer, Elleke. *Colonial and Postcolonial Literature*. New York: Oxford UP, 1995.

Bonner, Raymond. *At the Hand of Man: Peril and Hope for Africa's Wildlife*. New York: Knopf, 1993.

Brantlinger, Patrick. *Rule of Darkness: British Literature and Imperialism, 1830–1914*. Ithaca: Cornell UP, 1988.

Brockington, Dan. *Fortress Conservation: The Preservation of the Mkomazi Game Reserve, Tanzania*. Oxford: James Currey, 2002.

Brown, Coreen. *The Creative Vision of Bessie Head*. Madison, NJ: Fairleigh Dickinson UP, 2003.

Brown, Lloyd. "Creating New Worlds in Southern Africa: Bessie Head and the Question of Power." *Umoja* 3.1 (1979): 43–53.

Buell, Lawrence. *The Environmental Imagination: Thoreau, Nature Writing, and the Formation of American Culture*. Cambridge: Harvard UP, 1995.

———. *The Future of Environmental Criticism*. Oxford: Blackwell, 2005.

———. *Writing for an Endangered World*. Cambridge: Harvard UP, 2001.

Bullis, Connie. "Retalking Environmental Discourses from a Feminist Perspective: Radical Potential of Ecofeminism." *The Symbolic Earth*. Ed. James Cantrill and Christine Oravec. Lexington: U of Kentucky P, 1996. 123–48.

Byerley, Andrew. *Becoming Jinja: The Production of Space and Making of Place in an African Industrial Town*. Stockholm: Stockholm University Department of Human Geography, 2005.

Caminero-Santangelo, Byron. *African Fiction and Joseph Conrad: Reading Postcolonial Intertextuality*. Albany: State U of New York P, 2005.

Caminero-Santangelo, Byron, and Garth Myers, eds. *Environment at the Margins: Literary and Environmental Studies in Africa*. Athens: Ohio UP, 2011.

Carruthers, Jane. *The Kruger National Park: A Social and Political History*. Pietermaritzburg: Natal UP, 1995.

——. "Nationhood and National Parks: Comparative Examples from the Post Imperial Experience." *Ecology and Empire*. Ed. Tom Griffiths and Libby Robin. Seattle: U of Washington P, 1997. 125–38.

Chabal, Patrick. *Africa: The Politics of Suffering and Smiling*. London: Zed, 2009.

Chrisman, Laura. "Nationalism and Postcolonial Studies." *The Cambridge Companion to Postcolonial Literary Studies*. Ed. Neil Lazarus. Cambridge: Cambridge UP, 2004. 183–98.

Cilano, Cara, and Elizabeth DeLoughrey. "Against Authenticity: Global Knowledges and Postcolonial Ecocriticism." *ISLE: Interdisciplinary Studies in Literature and Environment* 14.1 (2007): 71–87.

Clayton, Cherry. "'A World Elsewhere': Bessie Head as Historian." *English in Africa* 15.1 (1988): 55–69.

Cock, Jacklyn, and David Fig. "The Impact of Globalisation on Environmental Politics in South Africa, 1990–2002." *African Sociological Review* 5.2 (2001): 15–35.

Coetzee, J. M. *White Writing: On the Culture of Letters in South Africa*. New Haven: Yale UP, 1988.

Conrad, Joseph. *Heart of Darkness*. Ed. Robert Kimbrough. New York: Norton, 1988.

Coupe, Laurence, ed. *The Green Studies Reader*. New York: Routledge, 2000.

Cowen, Michael and Robert Shenton. "The Invention of Development." *Power of Development*. Ed. Jonathan Crush. London: Routledge, 1995. 27–43.

Cronon, William. "The Trouble with Wilderness; or, Getting Back to the Wrong Nature." *Uncommon Ground: Rethinking the Human Place in Nature*. Ed. William Cronon. New York: Norton, 1996. 69–90.

Crosby, Alfred W. *Ecological Imperialism*. 2nd ed. Cambridge: Cambridge UP, 2004.

Curtin, Deane. *Environmental Ethics for a Postcolonial World*. Lanham: Rowman & Littlefield, 2005.

Debbane, Anne-Marie, and Roger Keil. "Multiple Disconnections: Environmental Justice and Urban Water in Canada and South Africa." *Space and Polity* 8.2 (2004): 209–25.

Dellier, Julien, and Sylvain Guyot. "The Fight for Land Rights versus Outsider's 'Appetites': Wild Coast Eco-Frontier Dynamics." *Rethinking the Wild Coast, South Africa: Eco-frontiers vs. livelihoods in Pondoland*. Ed. Sylvain Guyot and Julien Dellier. Saarbrucken, Germany: VDM Muller, 2009. 59–100.

DeLoughrey, Elizabeth, Renee Gosson, and George Handley. *Caribbean Literature and the Environment*. Charlottesville: U of Virginia P, 2005.

DeLoughrey, Elizabeth, and George Handley, eds. *Postcolonial Ecologies: Literatures of the Environment*. Oxford: Oxford UP, 2011.

Derman, Bill, Rie Odgaard, and Espen Sjaastad, eds. *Conflicts over Land and Water in Africa*. Oxford: James Currey, 2007.

Dixon, Robin. "Nigerian Oil Spills Have Created Ecological Disaster." *Los Angeles Times*. August 5, 2011. Web.

Douglas, Oronto, Dimieari von Kemedi, Ike Okonta, and Michael J. Watts. "Alienation and Militancy in the Niger Delta: Petroleum, Politics, and Democracy in Nigeria." *The Quest for Environmental Justice: Human Rights and the Politics of Pollution*. Ed. Robert D. Bullard. San Francisco: Sierra Club Books, 2005. 239–54.

Drayton, Richard. *Nature's Government: Science, Imperial Britain, and the "Improvement" of The World*. New Haven: Yale UP, 2000.

Duffy, Rosaleen. *Killing for Conservation: Wildlife Policy in Zimbabwe*. Oxford: James Currey, 2000.

Eichstaedt, Peter. *Pirate State*. Chicago: Lawrence Hill, 2010.

Egya, Sule Emmanuel. "The Aesthetic of Rage in Recent Nigerian Poetry in English: Olu Oguibe and Ifowodo Ogaga." *Matatu: Journal of African Culture and Society* 39.2 (2011): 99–114.

———. "Imagining the Beast: A Critique of Images of the Oppressor in Recent Nigerian Poetry in English." *Journal of Commonwealth Literature* 46.2 (2011): 345–58.

Eldridge, Michael. "Out of the Closet: Farah's *Secrets*." *Emerging Perspectives on Nuruddin Farah*. Ed. Derek Wright. Trenton, NJ: Africa World Press, 2002. 637–60.

Escobar, Arturo. "Place, Economy, and Culture in a Post-Development Era." *Places and Politics in an Age of Globalization*. Ed. Roxann Prazniak and Arif Dirlik. New York: Rowman & Littlefield, 2001. 193–217.

Fairhead, James, and Melissa Leach. *Misreading the African Landscape: Society and Ecology in a Forest-Savannah Mosaic*. Cambridge: Cambridge UP, 1996.

Fanon, Frantz. *The Wretched of the Earth*. New York: Grove, 1963.

Farah, Nuruddin. *Crossbones*. New York: Penguin, 2011.

———. *Secrets*. New York: Penguin, 1998.

———. *Yesterday, Tomorrow: Voices from the Somali Diaspora*. New York: Cassell, 2000.

Ferguson, James. *The Anti-Politics Machine: "Development," Depoliticization and Bureaucratic Power in Lesotho*. Cambridge: Cambridge UP, 1990.

———. *Global Shadows: Africa in the Neoliberal World Order*. Durham: Duke UP, 2006.

Fielding, Maureen. "Agriculture and Healing: Transforming Space, Transforming Trauma in Bessie Head's *When Rain Clouds Gather*." *Critical Essays on Bessie Head*. Ed. Maxine Sample. Westport, CT: Praeger, 2003. 11–24.

Finnegan, William. "How I Got the Story: A Collection of Reminiscences by a Polish Journalist on his 40-Year Career of Covering the Third-World." *New York Times Book Review*. May 27, 2001. 11.

Gadgil, Madhav, and Ramachandra Guha. *This Fissured Land: An Ecological History of India*. Berkeley: U of California P, 1993.

Garrard, Greg. *Ecocriticism*. New York: Routledge, 2004.

Gates, Henry Louis. *The Signifying Monkey*. Oxford: Oxford UP, 1988.

Gifford, Terry. *Pastoral*. London: Routledge, 1999.

Gikandi, Simon. *Reading Chinua Achebe*. Portsmouth, NH: Heinemann, 1991.

Glotfelty, Cheryll, and Harold Fromm, eds. *The Ecocriticism Reader: Landmarks in Literary Ecology.* Athens: U of Georgia P, 1996.

Gorak, Irene. "Libertine Pastoral: Nadine Gordimer's *The Conservationist.*" *Novel* 24 (1992): 241–56.

Gordimer, Nadine. *The Conservationist.* New York: Penguin, 1972.

———. *Get a Life.* New York: Penguin, 2005.

———. "Living in the Interregnum." *The Essential Gesture.* Ed. Stephen Clingman. New York: Penguin, 1988. 261–84.

———. *Writing and Being.* Cambridge: Harvard UP, 1995.

Graham, James. "From Exceptionalism to Social Ecology in Southern Africa: Isolation, Intimacy and Environment in Nadine Gordimer's *Get a Life.*" *Rerouting the Postcolonial: New Directions for the New Millennium.* Ed. Janet Wilson, Cristina Sandru, and Sarah Lawson Welsh. New York: Routledge, 2010. 194–205.

Griffiths, Tom. "Introduction." *Ecology and Empire: Environmental History of Settler Societies.* Ed. Tom Griffiths and Libby Robin. Seattle: U of Washington P, 1997. 1–16.

Grove, Richard. *Green Imperialism: Colonial Expansion, Tropical Island Edens and the Origin of Environmentalism, 1600–1800.* Cambridge: Cambridge UP, 1995.

Guha, Ramachandra. *Environmentalism: A Global History.* New York: Longman, 2000.

Guha, Ramachandra, and Joan Martinez-Alier. *Varieties of Environmentalism: Essays North and South.* London: Earthscan, 1997.

Hall, Stuart. "Cultural Identity and Diaspora." *Colonial Discourse and Post-Colonial Theory.* Ed. Patrick Williams and Laura Chrisman. New York: Columbia UP, 1994. 392–403.

Hallowes, David. "The Environment of Apartheid-Capitalism." *Unsustainable South Africa: Environment, Development, and Social Protest.* Ed. Patrick Bond. Pietermaritzburg: Natal UP, 2002. 25–50.

Hallowes, David, and Mark Butler. "Power, Poverty, and Marginalized Environments: A Conceptual Framework." *Environmental Justice in South Africa.* Ed. David A. McDonald. Athens: Ohio UP, 2002. 51–77.

Harvey, David. *Cosmopolitanism and the Geographies of Freedom.* New York: Columbia UP, 2009.

———. *Justice, Nature and the Geography of Difference.* Malden: Blackwell, 1996.

Head, Bessie. *When Rain Clouds Gather.* London: Heinemann, 1995. First published 1969.

Head, Dominic. "The (Im)possibility of Ecocriticism." *Writing the Environment: Ecocriticism and Literature.* Ed. Richard Kerridge and Neil Sammells. New York: Zed, 1998. 27–39.

Heise, Ursula K. "The Hitchhiker's Guide to Ecocriticism." *PMLA* 121.2 (March 2006): 503–16.

———. *Sense of Place and Sense of Planet.* Oxford: Oxford UP, 2008.

Heron, G. A. Introduction. *Song of Lawino and Song of Ocol.* By Okot p'Bitek. London: Heinemann, 1984. 1–33.

———. *The Poetry of Okot p'Bitek*. London: Heinemann, 1976.

Highfield, Jonathan. "'Relations with Food': Agriculture, Colonialism, and Food-ways in the Writing of Bessie Head." *Postcolonial Green: Environmental Politics and World Narratives*. Ed. Bonnie Roos and Alex Hunt. Charlottesville: U of Virginia P, 2010. 102–17.

Hitchcock, Peter. "Postcolonial Failure and the Politics of Nation." *South Atlantic Quarterly* 106.4 (2007): 727–52.

Huggan, Graham. "'Greening' Postcolonialism: Ecocritical Perspectives." *Modern Fiction Studies* 50.3 (2004) 701–33.

Huggan, Graham, and Helen Tiffin. "Green Postcolonialism." *Interventions: International Journal of Postcolonial Studies* 9.11 (2007) 1–11.

———. *Postcolonial Ecocriticism: Literature, Animals, Environment*. London: Routledge, 2010.

Hughes, David McDermott. "Whites Lost and Found: Immigration and Imagination in Savanna Africa." *Environment at the Margins: Literary and Environmental Studies in Africa*. Ed. Byron Caminero-Santangelo and Garth Myers. Athens: Ohio UP, 2011. 159–84.

Hulme, David, and Marshall Murphree, eds. *African Wildlife and Livelihoods*. Portsmouth: Heinemann, 2001.

Huxley, Elspeth. *The Flame Trees of Thika: Memories of an African Childhood*. New York: Penguin, 2000. First published 1959.

Hyden, Goran. "Governance and the Reconstitution of Political Order." *State, Conflict and Democracy in Africa*. Ed. Richard Joseph. Boulder: Lynne Rienner. 1999. 179–96.

Ibrahim, Huma. *Bessie Head: Subversive Identities in Exile*. Charlottesville: U of Virginia P, 1996.

Ifowodo, Ogaga. *The Oil Lamp*. Trenton, NJ: Africa World Press, 2005.

Izevbaye, Dan. "Chinua Achebe and the African Novel." *The Cambridge Companion to the Africa Novel*. Ed. F. Abiola Irele. Cambridge: Cambridge UP, 2009. 31–50.

Jacobs, J. U. "Zakes Mda's *The Heart of Redness*: The Novel as Umngqokolo." *Kunapipi: Journal of Post-Colonial Writing*. 24.1–2 (2002): 224–36.

Junger, Sebastian. "Blood Oil." *Vanity Fair* 49.2 (2007): 112.

Kalan, H., and B. Peek. "South African Perspectives on Transnational Environmental Justice Networks." *Power, Justice and the Environment: A Critical Appraisal of the Environmental Justice Movement*. Ed. D. N. Pellow and R. J. Brulle. Cambridge: MIT P, 2005. 253–63.

Kanogo, Tabitha. *Squatters and the Roots of Mau Mau, 1905–1963*. Nairobi: Heinemann, 1987.

Kaplan, Robert. "The Coming Anarchy: How Scarcity, Crime, Overpopulation, and Disease are Rapidly Destroying the Social Fabric of Our Planet." *Atlantic Monthly* 273.2 (February 1994): 44–76.

Kapuscinski, Ryszard. *The Shadow of the Sun*. New York: Random House, 2001.

Khan, Farieda. "The Roots of Environmental Racism and the Rise of Environmental Justice in the 1990s." *Environmental Justice in South Africa*. Ed. David A. McDonald. Athens: Ohio UP, 2002. 15–48.

Klopper, Dirk. "Between Nature and Culture: The Place of Prophecy in Zakes Mda's *The Heart of Redness*." *Current Writing: Text and Reception in Southern Africa* 20.2 (2008): 92–107.

Knipp, Thomas R. "Kenya's Literary Ladies and the Mythologizing of the White Highlands." *South Atlantic Review* 55.1 (1990): 1–16.

Kolodny, Annette. "Rethinking the 'Ecological Indian': A Penobscot Precursor." *ISLE: Interdisciplinary Studies in Literature and Environment* 14.1 (2007): 1–23.

Koyana, S. "Qolorha and the Dialogism of Place in Zakes Mda's *The Heart of Redness*." *Current Writing* 15.1 (2003): 51–62.

Krech, Shepard, III. *The Ecological Indian: Myth and History*. London: Norton, 1999.

Lacey, Marc. "To Fuel the Mideast's Grills, Somalia Smolders." *New York Times*. July 25, 2002: A4.

Laye, Camara. *The Dark Child*. Trans. James Kirkup and Ernest Jones. New York: Noonday, 1994.

Leach, Melissa, and Robin Mearns, eds. *The Lie of the Land: Challenging Received Wisdom on the African Environment*. Portsmouth, NH: Heinemann, 1996.

Leonard, L., and M. Pelling. "Mobilisation and Protest: Environmental Justice in Durban, South Africa." *Local Environment* 15.2 (2010): 137–51.

Leopold, Aldo. *A Sand County Almanac and Sketches Here and There*. New York: Oxford UP, 1989. First published 1949.

Lewis, Simon. *White Women Writers and Their African Inventions*. Gainesville: UP of Florida, 2003.

Loomba, Ania. *Colonialism/Postcolonialism*. 2nd ed. New York: Routledge, 2005.

Lundblad, Michael. "Malignant and Beneficent Fictions: Constructing Nature in Ecocriticism and Achebe's *Arrow of God*." *West Africa Review* 3.1 (2001): 1–21.

Maathai, Wangari. *The Challenge for Africa*. New York: Pantheon, 2009.

———. *Unbowed: A Memoir*. New York: Anchor, 2006.

Mackenzie, A. Fiona D. "Contested Ground: Colonial Narratives and the Kenyan Environment, 1920–1945." *Journal of Southern African Studies* 26.4 (2000): 697–718.

———. *Land, Ecology and Resistance in Kenya, 1880–1952*. Portsmouth, NH: Heinemann, 1998.

Martin, Julia. "New, with Added Ecology? Hippos, Forests, and Environmental Literacy." *Interdisciplinary Studies in Literature and Environment* 2.1 (1994): 1–11.

Martinez-Alier, Joan. *The Environmentalism of the Poor: A Study of Ecological Conflicts and Valuation*. Cheltenham, UK: Edward Elgar, 2002.

Massey, Doreen. "A Global Sense of Place." *Reading Human Geography*. Ed. Trevor Barnes and Derek Gregory. London: Arnold, 1997. 315–23.

Mbembe, Achille. *On the Postcolony*. Berkeley: U of California P, 2001.

McClintock, Anne. *Imperial Leather: Race, Gender and Sexuality in the Colonial Contest*. New York: Routledge, 1995.

McDonald, David A. "Environmental Racism and Neoliberal Disorder in South Africa." *The Quest for Environmental Justice: Human Rights and the Politics of Pollution.* Ed. Robert D. Bullard. San Francisco: Sierra Club Books, 2005. 255–78.

———. "What Is Environmental Justice?" *Environmental Justice in South Africa.* Ed. David A. McDonald. Athens: Ohio UP, 2002. 1–12.

McGinnis, Michael V. "A Rehearsal to Bioregionalism." *Bioregionalism.* Ed. Michael V. McGinnis. New York: Routledge, 1999. 1–9.

Mda, Zakes. *The Heart of Redness.* New York: Picador, 2000.

———. "A Response to 'Duplicity and Plagarism in Zakes Mda's *The Heart of* Redness' by Andrew Offenburger." *Research in African Literatures* 39.3 (2008): 200–3.

Miller, Christopher. *Theories of Africans: Francophone Literature and Anthropology in Africa.* Chicago: U of Chicago P, 1990.

Mitchell, Timothy. "The Object of Development: America's Egypt." *Power of Development.* Ed. Jonathan Crush. London: Routledge, 1995. 129–57.

Myers, Garth. *African Cities: Alternative Visions of Urban Theory and Practice.* London: Zed, 2011.

Neumann, Roderick. *Imposing Wilderness: Struggles over Livelihood and Nature Preservation in Africa.* Berkeley: U of California P, 1998.

———. *Making Political Ecology.* New York: Hodder Arnold, 2005.

Ngaboh-Smart, Francis. "*Secrets* and a New Civic Consciousness." *Research in African Literatures* 31.1 (2000): 129–36.

Ngũgĩ wa Thiong'o. *A Grain of Wheat.* London, Heinemann, 1986. First published 1967.

———. *Moving the Centre.* Portsmouth: Heinemann, 1993.

———. *Penpoints, Gunpoints, and Dreams: Towards a Critical Theory of the Arts and the State in Africa.* Oxford: Clarendon, 1998.

———. *Petals of Blood.* New York: Penguin, 1977.

Nicholls, Brendon. "The Landscape of Insurgency: Mau Mau, Ngugi wa Thiong'o and Gender." *Landscape and Empire, 1770–2000.* Ed. Glenn Hooper. Aldershot: Ashgate, 2005. 177–94.

"Nigeria Ogoniland Oil Clean-up 'Could Take 30 Years.'" *BBC News.* August 4, 2011. Web.

Nixon, Rob. "Environmentalism and Postcolonialism." *Postcolonial Studies and Beyond.* Ed. Ania Loomba, Suvir Kaul, Matti Bunzl, Antoinette Burton, and Jed Esty. Durham: Duke UP, 2005. 233–51.

———. *Homelands, Harlem and Hollywood.* New York: Routledge, 1994.

———. "Pipe Dreams: Ken Saro-Wiwa, Environmental Justice, and Microminority Rights." *Ken Saro-Wiwa: Writer and Political Activist.* Ed. Craig McLuckie and Aubrey McPhail. Boulder: Lynne Rienner, 2000. 109–25.

———. *Slow Violence and the Environmentalism of the Poor.* Cambridge: Harvard UP, 2011.

Nkosi, Lewis. *Home and Exile.* New York: Longman, 1983.

O'Brien, Susie. "Articulating a World of Difference: Ecocriticism, Postcolonialism, and Globalization." *Canadian Literature* 170–71 (2001): 140–58.

———. "'Back to the World': Reading Ecocriticism in a Postcolonial Context." *Five Emus to the King of Siam*. Ed. Helen Tiffin. Amsterdam: Rodopi, 2007. 177–99.

Oelofse, C., D. Scott, G. Oelofse, and J. Houghton. "Shifts within Ecological Modernisation in South Africa: Deliberation, Innovation, and Institutional Opportunities." *Local Environment* 11.1 (2006): 61–78.

Offenburger, Andrew. "Duplicity and Plagiarism in Zakes Mda's *The Heart of Redness*." *Research in African Literatures* 39.3 (2008): 164–99.

Ojaide, Tanure. *Contemporary African Literature: New Approaches*. Durham: Carolina Academic Press, 2012.

———. *Delta Blues and Home Songs*. Ibadan: Kraft Books, 1998.

———. "Foreword." *Eco-Critical Literature: Regreening African Landscapes*. Ed. Ogaga Okuyade. New York: African Heritage Press, 2013. v–viii.

———. *The Tale of the Harmattan*. Cape Town: Kwela, 2007.

Okonta, Ike, and Oronto Douglas. *Where Vultures Feast: Shell, Human Rights, and Oil in the Niger Delta*. San Francisco: Sierra Club, 2001.

Okot p'Bitek. *Song of Lawino and Song of Ocol*. Ed. G. A. Heron. London: Heinemann, 1984.

Okpewho, Isidore. *African Oral Literature: Backgrounds, Character, and Continuity*. Bloomington: Indiana UP, 1992.

Okuyade, Ogaga, ed. *Eco-Critical Literature: Regreening African Landscapes*. New York: African Heritage Press, 2013.

Osaghae, Eghosa E. "The Ogoni Uprising: Oil Politics, Minority Agitations and the Future of the Nigerian State." *African Affairs* 94.376 (1995): 325–44.

Parry, Benita. "Resistance Theory/Theorizing Resistance; or, Two Cheers for Nativism." *Colonial Discourse/Postcolonial Theory*. Ed. Francis Barker, Peter Hulme, and Margaret Iversen. Manchester: Manchester UP, 1994. 172–96.

Paton, Alan. *Cry, the Beloved Country*. New York: Scribner, 2003. First published 1948.

Peet, Richard, Paul Robbins, and Michael J. Watts, eds. *Global Political Ecology*. New York: Routledge, 2011.

Peires, J. *The Dead Will Arise: Nongqawuse and the Great Xhosa Cattle Killing Movement of 1856–7*. Bloomington: Indiana UP, 1989.

Peluso, Nancy Lee, and Michael Watts. "Violent Environments." *Violent Environments*. Ed. Nancy Lee Peluso and Michael Watts. Ithaca: Cornell UP, 2005. 3–38.

Perfecto, Ivette, John Vandermeer, and Angus Wright. *Nature's Matrix: Linking Agriculture, Conservation, and Food Sovereignty*. London: Earthscan, 2009.

Pilkington, Ed. "Shell Pays Out $15.5m over Saro-Wiwa Killing." *Guardian*. June 9, 2009. Web.

Plant, Judith. *Healing the Wounds: The Promise of Ecofeminism*. London: Green Print, 1989.

Plumwood, Val. *Environmental Culture: The Ecological Crisis of Reason*. London, Routledge, 2002.

———. *Feminism and the Mastery of Nature*. London: Routledge, 1993.

Pratt, Mary Louise. *Imperial Eyes: Travel Writing and Transculturation*. New York: Routledge, 1992.

Quayson, Ato. *Calibrations*. Minneapolis: U of Minnesota P, 2003.

Raji, Wumi. "Oil Resources, Hegemonic Politics and the Struggle for Re-inventing Post-colonial Nigeria." *Ogoni's Agonies: Ken Saro-Wiwa and the Crisis of Nigeria*. Ed. Abdul Rasheed Na'Allah. Trenton, NJ: Africa World Press, 1998. 109–20.

Ribot, Jesse, and Phil Oyono. "The Politics of Decentralization." *Towards a New Map of Africa*. Ed. Ben Wisner, Camilla Toulmin, and Rutendo Chitiga. London: Earthscan, 2006. 205–28.

Robin, Libby. "Ecology: A Science of Empire." *Ecology and Empire: Environmental History of Settler Societies*. Edinburgh: Keele UP, 1997. 63–75.

Ross, Andrew. *The Chicago Gangster Theory of Life: Nature's Debt to Society*. London: Verso, 1994.

Rowell, Andrew, James Marriott, and Lorne Stockman. *The Next Gulf: London, Washington and Oil Conflict in Nigeria*. London: Constable and Robertson, 2005.

Sample, Maxine. "Space: An Experiential Perspective: Bessie Head's *When Rain Clouds Gather*." *Critical Essays on Bessie Head*. Ed. Maxine Sample. Westport, CT: Praeger, 2003. 25–45.

Saro-Wiwa, Ken. *Genocide in Nigeria: The Ogoni Tragedy*. London: Saros International, 1992.

———. *A Month and a Day: A Detention Diary*. Oxfordshire: Ayebia, 1995.

Schlosberg, David. *Defining Environmental Justice: Theories, Movements and Nature*. Oxford: Oxford UP, 2007.

Schofield, Clive. "Plundered Waters: Somalia's Maritime Resource Insecurity." *Crucible for Survival: Environmental Security and Justice in the Indian Ocean*. Ed. Timothy Doyle and Melissa Risely. New Brunswick: Rutgers UP, 2008. 102–15.

Shetler, Jan Bender. *Imagining Serengeti: A History of Landscape Memory in Tanzania from Earliest Times to the Present*. Athens: Ohio UP, 2007.

Simukonda, Navy, and Mcebisi Kraai. "The Wild Coast: the Contested Territory." *Rethinking the Wild Coast, South Africa: Eco-frontiers vs. Livelihoods in Pondoland*. Ed. Sylvain Guyot and Julien Dellier. Saarbrucken, Germany: VDM Muller, 2009. 33–57.

Slaymaker, William. "Echoing the Other(s): The Call of Global Green and Black African Responses." *PMLA* 116 (2001): 129–44.

Smith, David. "Shell Accused of Fuelling Violence in Nigeria by Paying Rival Militant Gangs." *Guardian*. October 3, 2011. Web.

Smith, Neil. "The Production of Nature." *FutureNatural: Nature/Science/Culture*. Ed. George Robertson, Melinda Mash, Lisa Tickner, Jon Bird, Barry Curtis, and Tim Putnam. New York: Routledge, 1996. 35–54.

Soper, Kate. *What Is Nature? Culture, Politics and the Non-Human*. Cambridge: Blackwell, 1995.

Soyinka, Wole. "From a Common Back Cloth: A Reassessment of the African Literary Image." *American Scholar* 23.3 (1963): 387–96.

———. *The Open Sore of a Continent*. New York: Oxford UP, 1996.

Spivak, Gayatri. "Can the Subaltern Speak?" *Marxism and the Interpretation of Culture*. Ed. Cary Nelson and Lawrence Grossberg. Basingstoke: Macmillan Education, 1988. 271–313.

———. "The Rani of Sirmur: An Essay in Reading the Archives." *History and Theory* 24.3 (1985): 247–72.

Spurr, David. *The Rhetoric of Empire: Colonial Discourse in Journalism, Travel Writing and Imperial Administration*. Durham: Duke UP, 1993.

Stepan, Nancy Leys. *Picturing Tropical Nature*. Ithaca: Cornell UP, 2001.

Stratton, Florence. *Contemporary African Literature and the Politics of Gender*. London: Routledge, 1994.

Sturgeon, Noel. *Environmentalism in Popular Culture: Gender, Race, Sexuality, and the Politics of the Natural*. Tuscon: U of Arizona P, 2009.

Sugnet, Charles. "Nuruddin Farah's *Maps*: Deterritorialization and 'the Postmodern.'" *World Literature Today* 72.4 (1998): 739–46.

Sze, Julie, and J. K. London. "Environmental Justice at the Crossroads." *Sociology Compass* 2.4 (2008): 1331–54.

Theroux, Paul. *Dark Star Safari: Overland from Cairo to Cape Town*. New York: Houghton Mifflin, 2003.

———. *The Last Train to Zona Verde: My Ultimate African Safari*. New York: Houghton Mifflin, 2013.

Thomashow, Mitchell. *Bringing the Biosphere Home*. Cambridge: MIT P, 2002.

———. "Toward a Cosmopolitan Bioregionalism." *Bioregionalism*. Ed. Michael Vincent McGinnis. London: Routledge, 1999. 121–31.

Tiffin, Helen, ed. *Five Emus to the King of Siam: Environment and Empire*. Amsterdam: Rodopi, 2007.

Vital, Anthony. "'Another Kind of Combat in the Bush': *Get a Life* and Gordimer's Critique of Ecology in a Globalized World." *English in Africa* 35 (2008): 89–118.

———. "Situating Ecology in Recent South African Fiction: J. M. Coetzee's *The Lives of Animals* and Zakes Mda's *The Heart of Redness*." *Journal of Southern African Studies* 31 (2005): 297–313.

———. "Toward an African Ecocriticism: Postcolonialism, Ecology and *Life and Times of Michael K*." *Research in African Literatures* 39.1 (2008): 87–106.

———. "Waste and Postcolonial History: An Ecocritical Reading of J. M. Coetzee's *Age of Iron*." *Environment at the Margins: Literary and Environmental Studies in Africa*. Ed. Byron Caminero-Santangelo and Garth Myers. Athens: Ohio UP, 2011. 185–212.

Vital, Anthony, and Hans-Georg Erney, eds. *Postcolonial Studies and Ecocriticism*. Special issue of *Journal of Commonwealth and Postcolonial Studies* 13–14 (2007).

Wagner, Kathrin. *Rereading Nadine Gordimer*. Bloomington: Indiana UP, 1994.

Walker, Gordon. *Environmental Justice: Concepts, Evidence and Politics*. New York: Routledge, 2012.

Ward, David. *Chronicles of Darkness*. New York: Routledge, 1989.

Warren, Karen J., ed. *Ecological Feminist Philosophies.* Bloomington: Indiana UP, 1996.

Watts, Michael. "Petro-Violence: Community, Extraction, and Political Ecology of a Mythic Commodity." *Violent Environments.* Ed. Nancy Lee Peluso and Michael Watts. Ithaca: Cornell UP, 2001. 189–212.

———. "Sweet and Sour." *Curse of the Black Gold: 50 Years of Oil the Niger Delta.* Ed. Michael Watts. New York: Powerhouse, 2007. 36–47.

———. "Violent Environments: Petroleum Conflict and the Political Ecology of Rule in the Niger Delta, Nigeria." *Liberation Ecologies: Environment, Development, Social Movements.* Ed. Richard Peet and Michael Watts. New York: Routledge, 2004. 273–98.

———. "Visions of Excess: African Development in an Age of Market Idolatry." *Transition* 51 (1991): 124–41.

Watts, Michael, and Richard Peet. "Liberating Political Ecology." *Liberation Ecologies: Environment, Development, Social Movements.* Ed. Richard Peet and Michael Watts. New York: Routledge, 1996. 3–47.

Wenzel, Jennifer. *Bulletproof: Afterlives of Anticolonial Prophecy in South Africa and Beyond.* Chicago: U of Chicago P, 2009.

Williams, Glyn, and Emma Mawdsley. "Postcolonial Environmental Justice: Government and Governance in India." *Geoforum* 37.5 (2006): 660–70.

Wiwa, Ken. *In the Shadow of a Saint.* New York: Doubleday, 2000.

Wolff, Richard D. *The Economics of Colonialism: Britain and Kenya, 1870–1930.* New Haven: Yale UP, 1994.

Woodward, Wendy. "Laughing Back at the Kingfisher: Zakes Mda's *The Heart of Redness* and Postcolonial Humor." *Cheeky Fictions: Laughter and the Postcolonial.* Ed. Susanne Reichl and Mark Stein. Amsterdam: Rodopi, 2005. 287–99.

Worster, Donald. *The Wealth of Nature: Environmental History and the Ecological Imagination.* Oxford: Oxford UP, 1993.

Wright, Derek. "Mapping Farah's Fiction: The Postmodern Landscapes." *Emerging Perspectives on Nuruddin Farah.* Ed. Derek Wright. Trenton, NJ: Africa World Press, 2002. 95–129.

Wright, Laura. *Wilderness into Civilized Shapes: Reading the Postcolonial Environment.* Athens: U of Georgia P, 2010.

Wylie, Dan, ed. *Toxic Belonging? Identity and Ecology in Southern Africa.* Newcastle upon Tyne: Cambridge Scholars Publishing, 2008.

index